The Institute of Biology's
Studies in Biology no. 12

Plant

Breeding

by William J. C. Lawrence

O.B.E., V.M.H., F.I.Biol.
Formerly Head of Department of Physiology and Plant Culture, John Innes Institute

New York·

CRANE, RUSSAK & COMPANY, INC.
347 Madison Avenue
New York, New York 10017

First published 1968

First published in·
the United States of America in 1968

First published in Great Britain by
Edward Arnold (Publishers) Ltd.

Library af Congress Catalog Card Number: 68–55288

Printed in Great Britain by
William Clowes and Sons Ltd, London and Beccles

General Preface to the Series

It is no longer possible for one textbook to cover the whole field of Biology and to remain sufficiently up to date. At the same time students at school, and indeed those in their first year at universities, must be contemporary in their biological outlook and know where the most important developments are taking place.

The Biological Education Committee, set up jointly by the Royal Society and the Institute of Biology, is sponsoring, therefore, the production of a series of booklets dealing with limited biological topics in which recent progress has been most rapid and important.

A feature of the series is that the booklets indicate as clearly as possible the methods that have been employed in elucidating the problems with which they deal. There are suggestions for practical work for the student which should form a sound scientific basis for his understanding.

1967

INSTITUTE OF BIOLOGY
41 Queen's Gate
London, S.W.7.

Preface

This booklet is intended to serve as an introduction to the subject of plant breeding and to indicate the genetical principles on which the subject is based. The future welfare, indeed survival, of man depends largely on his ability to understand and manipulate the genetic mechanisms basic to the evolution, domestication and improvement of plants. With more than two-thirds of the world's population undernourished, the development of improved, high-yielding crop plants, adapted to many specific environments and requirements, is of great urgency, and must go hand in hand with the spread of advanced agricultural practices.

September, 1967

W.J.C.L.

Contents

Origins and Development of
Cultivated Plants

1.1 Selection and hybridization

Plant improvement by man began 9,000 years ago. Prior to this time he subsisted by hunting and fishing, supplementing his spoils by such wild roots, herbs, fruits and seeds as he found during his wanderings. It has been estimated that it took sixteen square miles of territory to support a single hunter and as a result men and their families were widely dispersed.

In these homeless wanderings pre-agricultural man must have gained an intimate knowledge of every useful (and harmful) plant in his domain (HELBAEK, 1959). Eventually Neolithic man learnt that seeds put into the soil at a certain time gave rise to similar seed-producing plants. This discovery most likely occurred when spilled seeds germinated on early man's nitrogen-rich middens. These dung heaps were made everywhere man halted for a while: thus all unconsciously he selected and bred quick-growing weeds which he carried about from place to place and consequently the weeds had excellent opportunities for crossing with other varieties of their kind to give improved sorts, or 'dung hill super-weeds' (ANDERSON, 1949).

In due course, more or less permanent, though primitive, settlement became possible, especially in regions where the profusion of game, grains and fruits and an adequate climate and water supply favoured the deliberate sowing of seeds and annual harvesting of crops. The evidence available suggests that the uplands of Asia Minor comprised one such favourable region which, indeed, was probably the cradle of Western civilization.

The conditions obtaining throughout man's first domestication of plants during the Neolithic revolution are highly relevant to the complex problems of scientific plant breeding in the present food-hungry world and will be for far into the future. The reasons for this are briefly outlined below.

Gene (and chromosome) mutation may be assumed to have proceeded at a more or less steady rate age after age. Gene mutation provides the basic material for natural selection. Now the great majority of gene mutations are deleterious and recessive in character; therefore, in so far as even a few of them are ultimately to become established by adaptation to the internal (balanced genotype) and external (climate, soils, etc.) plant environments, they must be exposed to the forces of selection. This is most likely to occur in populations that are genetically heterogenous, i.e. in heterozygous individuals capable of carrying and segregating recessive characters. In the wild, the great majority of plants are outbreeders (cross-pollinated) and hence wild populations are characteristically maintained in a heterozygous state.

As we have seen, the early domestication of plants in Neolithic times resulted in the bringing together of many wild species and varieties, providing a rich source of variation from which deliberate selection could be made in the light of slowly accumulated observation and experience. Man's early 'crops' therefore were large stores of genetic variability, concentrated locally from the dispersed store of variability in the wild.

At first, the isolation imposed by tribal, ethnic and topographical boundaries resulted in man having to subsist on a limited number of cultigens which differed a little genetically from those of neighbouring tribes. Later, as the consequence of war and trade, these numerous tribal cultigens became more widespread with further opportunities for hybridization and selection.

Eventually, as horses, camels and ships became increasingly available for transport purposes, this variability was still further added to by the introduction of seeds, roots and bulbs from more and more distant places. In this way began the first great population expansion of the human species, due to the increased productive capacity of plants, land and labour.

This sequence of events, first in the conservation and then extension of genetic variability, had another and opposite effect. Selection of improved varieties led gradually to the elimination of the less advantageous ones they displaced. For example, there is good evidence that the wild progenitors of wheat and barley from which domesticated forms were selected had a brittle rachis (stem of ear) which, when mature, efficiently shattered and dispersed the grains. A mutation giving a rachis that did not shatter but held the grains in a compact mass would be a great advantage for harvesting under cultivation but would rapidly die out in wild populations because of its less efficient seed dispersal. Selection by man of the plants with non-brittle rachis would quickly eliminate the wild forms in his crops.

Over many centuries plant improvement by selection of this kind must have not only reduced the number of different variants in each cultivated species but also diminished their capacity for variation, hence also the effectiveness of selection, a process which has reached an extreme form in many of the highly domesticated plants and crops of today.

In summary, the transport of plants by man, their accumulation and hybridization in definite areas, all involve gene transfer. Often, genes of advantage in one environment would be at a disadvantage in another; nevertheless slowly, even though rarely, new combinations of genes would arise adapted to new genomes (a genome is the basic haploid set of chromosomes in a species) and new external environments. This process of gene transfer and recombination under selection would have been extremely slow in Neolithic times, yet it greatly hastened the rise of man from a state of savagery to one of civilization. Man had accidentally set in motion and become involved in the genetic forces of evolution acting upon plants (and animals) under domestication (MANGELSDORF, 1965).

Thousands of years later, exploration, plant collection, easy and rapid transport of plant material by ships, trains, motor vehicles and aircraft were to increase enormously the processes set in motion by Neolithic man. By the nineteenth century plant material was being collected in quantity from all continents by European and North American plant hunters. The coming of the twentieth century witnessed the establishment of official plant collections on a large scale, e.g. 12,000 wheat and barley variants maintained by the USA world collection of these cereals, some 4,000 variants as tubers and seeds in the Commonwealth Potato Collection, and some 10,000 rice variants at the International Rice Research Institute in the Philippines.

Turning now to domesticated plants, the number of which exceeds 1,500 economic species with innumerable cultivars, we may distinguish three main aspects of their origin and evolution: (a) Mendelian segregation and recombination, (b) interspecific hybridization and (c) polyploidy.

1.2 Mendelian segregation and recombination

In the domestication of plants by man, many improvements have been made by simple selection of variants within a single species, e.g. tomato, cucumber, barley, rice, maize, flax. A character depending on a single spontaneous mutation within a species has often been directly selected to give a new variety. A fully documented example of this is provided by the history of the sweet pea ($2n=14$). Prior to 1900, all varieties had plain standards (upright petals) and flowers of moderate size. Then a mutant form with waved standards and larger flowers appeared in four localities, 1900–1902. In three of these the mutant form appeared in the variety Prima Donna; in the fourth locality the mutant appeared in seedlings from a cross with Prima Donna. The mutant form with waved standards was selected because of its attractiveness, at once bred true, and was named Countess Spencer.

The Spencer character, which is determined by a single recessive gene (h) which must have arisen in Prima Donna (HH), was hidden in a heterozygous condition in that variety for one generation and then segregated out in the homozygous recessive form (hh) in stocks probably derived from one and the same source of Prima Donna. Most sweet peas today have waved standards and have come from the one recessive mutation.

In contrast with the somewhat dramatic debut of the Spencer sweet pea is that of the Shirley strain of poppies, with white edges to their petals and of different colours. In 1880, in a patch of the common wild poppy, *Papaver rhoeas* ($2n=14$), with red petals having a central black blotch, one solitary flower was found with a very narrow white edging to the petals. Seed saved from this flower gave some 200 plants of which 4 to 5 were white-edged. By continued selection of the best variants, a range of colours was obtained from red to white, together with golden or white stamens, anthers and pollen, and a white blotch to the base of the petals. The coloured forms had white edges to the petals.

Subsequent genetic research showed that the white edge arose from the mutation of a recessive gene w to the dominant W, while the white blotch and yellow anthers, etc., arose from the mutation of a dominant gene B to the recessive b. Continued selection of the major characters must have also resulted in the unconscious selection of modifying genes that increased the expression of the major ones.

Although selection of single gene mutations has played a large part in the improvement of cultivated plants, it is only rarely that a sequence of such mutations in a given species has been authentically recorded. The development of the sweet pea from the wild form during some 200 years provides an almost unique example, partly because of its interest to gardeners who recorded 'sports', mainly due to its being an inbreeding diploid in which mutations of ornamental interest can be easily discerned and 'fixed'

Table 1   History of gene mutation in the sweet pea.

Character	Gene	Date of first appearance	Original mutant variety or source of origin
Wild sweet pea	—	Intro. 1699	—
White flowers	g_1 or f_1	1718	—
Red standards	a_1	1731 (?37)	Painted Lady
'Black'	—	1793	—
'Scarlet'	—	1793	—
Picotee	d_5	1860	—
Hooded	a_3	1886	Waverley
Dilute	g_3	1890	Countess of Radnor
Cupid	e	1893	ex Emily Henderson
Dull	d_2	1899	Navy Blue
Round pollen	a_2	Before 1900 ?	Emily Henderson
Waved standards	h	1900	Countess Spencer
Sterile anthers	b_2	1903	ex Emily Henderson × Lady Penzance
Copper	g_2	1905	Cambridge
Marbled	F_1	1905	Helen Pierce
Bush	f_2	Before 1905	—
Cretin	b_3	1907	Cambridge
Acacia	d_1	1908	Mr. W. J. Unwin
Smooth	d_4	1912	Mr. T. A. Dipnall

The characters with the same gene letter are linked in inheritance

(p. 21). The history of mutation in the sweet pea is given in Table 1 and illustrates what must be a commonplace phenomenon in many comparatively unrecorded instances in other species.

The genetic diversity produced by gene mutation has been further augmented by deliberate hybridization, and the subsequent gene recombinations have enabled selection to be made of new varieties for different purposes and adapted to different conditions. In the broad, this intraspecific variation may be termed Mendelian segregation and recombination and until the dawn of this century by far the greater number of advances in plant improvement had been made by simple selection of this kind, e.g. among horticultural plants, round- as opposed to rib-fruited tomatoes, aberrant forms of the wild cabbage which were the progenitors of cabbage, Brussels sprouts, kale, etc., apples such as Cox's Orange Pippin and Bramley's Seedling, the Shirley poppy, double-flowered cultivars.

Among crop plants, alfalfa, asparagus, beans, beets, carrots, celery, lettuce, lupin, onion, radish, soybean and tomato are examples in which improvement has been mainly, perhaps solely, by selection. A particularly well recorded case is provided by sugar-beet, developed from the mangelwurzel grown for long years as a fodder crop in Europe. When beets were first grown for the manufacture of sugar about 150 years ago, the average sucrose content was little more than 7 per cent. This figure had increased to 10 per cent by 1870 and, following the introduction of polariscopic analysis of the juice as a guide to selection, to 14 per cent by 1890, and to 16 per cent by about 1910.

1.3 Interspecific hybridization

Interspecific hybridization has often led to a great increase in the number of variants in segregating generations, from which selection has preserved forms valuable to man. This is seen *par excellence* in a great many ornamental plants; rhododendrons, roses, dahlias, chrysanthemums, poppies, violets, gladioli, lilies, freesias, orchids, each of which embraces numerous varieties. Even when interspecific hybrids are sterile they may find a place in gardens, since multiplication by vegetative means is usually possible. Interspecific hybridization has also yielded cultivars in apple, grape, tomato (recently), loganberry, maize and rice, and to a small extent, forest trees.

Two examples of the part played by interspecific hybridization concern the strawberry and wheat. Up to the end of the sixteenth century the only species of strawberry grown in gardens in Europe were the diploid *Fragaria moschata* syn. *vesca* ($2n = 14$) and the hexaploid *F. elatior* ($2n = 42$). Very early in the seventeenth century the octoploid *F. virginiana* ($2n = 56$) was introduced to Europe from the Eastern seaboard of USA, but although 20 to 30 varieties were cultivated in gardens by the end of the eighteenth century, no notable progress had been made in the development of this or the other two species, probably because they lacked the necessary variability. Due to their different chromosome numbers, no hybridization was possible between *F. moschata*, *F. elatior*, and *F. virginiana*.

Early in the eighteenth century a fourth species was introduced into

Europe, the octoploid *F. chiloensis* ($2n = 56$) from the western coast of Chile. This species bears large fruits and is very distinct from the others. Because of breeding and cropping difficulties caused by dioecy and poor fruit quality, no progress was made in the improvement of this species.

In the early part of the nineteenth century English breeders made crosses between *F. virginiana* and *F. chiloensis*, selecting for hermaphrodite flowers, and fruits of larger size and better quality. This was the origin of the modern commercial strawberry, which came from the crossing of two species, having the same chromosome number, brought together in Europe from two widely separated parts of the world.

The history of the origin of the cultivated strawberry is fully recorded. In the case of wheat, its origin in remote times has been deduced largely from cytogenetical studies. The full story of the unravelling of the evolution of the bread wheat, *Triticum aestivum* (RILEY, 1965), is remarkable in demonstrating the ability of modern cytogenetic studies to trace the phylogeny of long-domesticated crops. The following account merely outlines the conclusions of this work.

The genus *Triticum* consists of a series of species which are either diploids ($2n = 14$), tetraploids ($2n = 28$) or hexaploids ($2n = 42$), hence the basic chromosome number of the genus is 7. The relationship of the diploid, tetraploid and hexaploid species was determined mainly from comparison of the pairing behaviour of the chromosomes in hybrids between these species. This showed that the species could be differentiated by their genomes, conventionally symbolized by the letters *A*, *B* and *D*: diploids *AA*, tetraploids *AABB* and hexaploids *AABBDD*.

The tetraploids arose from hybridization between diploid species contributing the *A* and *B* genomes respectively, followed by chromosome doubling to give the *AABB* forms. Later, hybridization took place

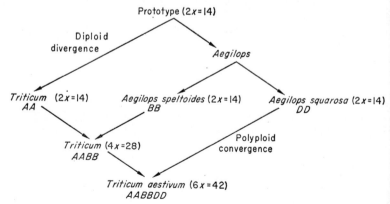

Fig. 1-1 The phylogeny of the bread wheat, *T. aestivum*. (Riley, R. from *Crop Plant Evolution* ed. Hutchinson, J. B., 1965. Cambridge University Press, London.)

between the tetraploid wheat, *AABB*, and a third diploid species, the *D* donor, followed by chromosome doubling to give the hexaploids *AABBDD* The species involved in the evolution of the bread wheat are shown in Fig. 1–1. This figure also demonstrates how first there was divergence in the evolution of the diploid species followed later by convergence in the evolution of the polyploid species, processes which must have occurred many times in many genera in the evolution of the flowering plants (cf. dahlia).

It will be noted that in strawberry and wheat, polyploidy has accompanied interspecific hybridization at some stage or other. This is almost always the case in the hybridization, natural or artificial, of distinct species, and the following section provides further examples of this.

1.4   Polyploidy

Polyploidy has, time and again, been a potent factor in the evolution of plants, and it has been estimated that over 40 per cent of the dicotyledons and nearly 60 per cent of the monocotyledons and at least a third of domesticated species (70 per cent of forage grasses) are polyploids. Autopolyploidy in which identical, or almost identical, genomes have been replicated (Fig. 1–2) has given many of our ornamental and horticultural varieties, especially because autopolyploids often bear larger flowers, fruits and leaves than their diploid counterparts. Autopolyploids, however, characteristically show reduced fertility due to aberrant chromosome behaviour at meiosis.

For example, competition in the pairing of the four identical chromosomes of each type in an autotetraploid reduces the frequency of chiasma formation between any two of the four chromosomes and this often results in one or more of them remaining unpaired, giving 1 trivalent plus 1 univalent, 1 bivalent plus 2 univalents, or 2 bivalents.

Further, should multivalent formation occur, the arrangement of the multivalents at metaphase may be of such a kind as to result in irregular distribution of the chromosomes to the daughter nuclei. Four chromosomes connected end to end to form a ring quadrivalent will probably disjoin into 2 plus 2 whatever their orientation may be at metaphase. In contrast, a chain of four chromosomes may disjoin into 2 plus 2, 3 plus 1, or all four may pass to one pole. The end result of this irregular distribution is many gametes with variable chromosome numbers. These are non-functional and fertility is proportionately reduced. Because of this infertility the number of genuine autopolyploids that have become established in nature is probably relatively small. The drawback of reduced fertility in the autopolyploids is compensated by the fact that many of them can be propagated asexually, e.g. apples, pears, potato, peanut, coffee.

Allopolyploidy, the result of hybridization between species followed by doubling of the chromosome number (Fig. 1–2), accounts for as many as half of cultivated plants. Allopolyploids combine the potential of two species in one and are usually fertile since the two genomes contributed to

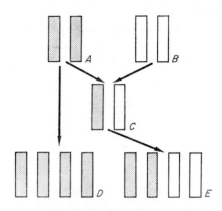

Fig. 1–2 Diagram showing the origin and constitution of the chromosome complements of auto- and allotetraploids. *A* and *B*, diploid species; *C*, diploid hybrid from *A* × *B*; *D*, autotetraploid derived from doubling of the chromosome sets in *A*; *E*, allotetraploid derived from the doubling of the chromosome sets in *C*. (CRANE, M. B. and LAWRENCE, W. J. C., 1952. *Genetics of Garden Plants*, Macmillan, London.)

the more or less sterile hybrid (in which the chromosome sets are too different to pair successfully) on doubling have identical partners and normal pairing is restored. Thus, in contrast to the autotetraploid, chromosome pairing in the allotetraploid results in bivalent formation and these bivalents regularly disjoin at meiosis to give fertile gametes with the haploid number of chromosomes. Indeed, in terms of chromosome pairing and the inheritance of characters, the true allotetraploid is a functional diploid.

Allopolyploids, if successful in competition with their parents, constitute new species (cf. wheat and dahlia). Allopolyploidy, therefore, is a mechanism that can break down the genetic barriers between species, combine their differences, and provide a wider scope for selection among the off-spring (cf. dahlia, p. 9). Without allopolyploidy, many species hybrids could never have perpetuated themselves. Examples of allopolyploids are tobacco, domestic plum, dahlia, sugar-cane, wheat, coffee, oats.

Among species of economic interest the origin of the following illustrate evolution by polyploidy. Commercial tobacco, *Nicotiana tabacum* ($2n = 48$) does not, and probably never did, occur as a natural species. Related species have 24 chromosomes and from 2 of these species, *N. sylvestris* and *N. tomentosum*, a sterile hybrid may be obtained which, on doubling the chromosome number, becomes moderately fertile and is equivalent in many respects to commercial tobacco, which originated many centuries ago.

The hexaploid domestic plum *Prunus domestica*, ($2n=48$) probably arose centuries ago from hybridization of *P. divaricata* ($2n=16$) and *P. spinosa* ($2n=32$) to give a sterile triploid hybrid, followed by chromosome doubling and the restoration of fertility. The ground colour of fruits in *divaricata* is yellow and the anthocyanin pigment red: in *spinosa* the ground colour is green and the anthocyanin blue. All four colours occur in *domestica*, together with their combinations. In the forests of the North Caucasus, *spinosa* and *divaricata* hybridize freely and a hexaploid has been bred from these two species (RYBIN, 1936). *Prunus domestica* has probably arisen on many occasions in prehistoric times.

The origin of the garden dahlia, *D. pinnata*, has been investigated in considerable detail, cytologically, genetically and chemically. All species cytologically examined are tetraploids ($2n=32$, one $2n=36$) except the garden dahlia which has double the number of the others, viz $2n=64$. All the species except the garden dahlia fall into two classes for flower colour: Group 1, pale magenta, ground colour ivory, and Group 2, deep scarlet-orange, ground colour yellow. All these colours, together with their combinations, occur in the garden dahlia (LAWRENCE and SCOTT-MON-CRIEFF, 1935).

Inheritance in the garden dahlia is tetrasomic (see p. 28) and shows that it is a double autotetraploid, i.e. hybrid octoploid, which combines the features of the other species. The genes for pale anthocyanin and ivory pigment have been derived via Group 1 tetraploids and the genes for deep anthocyanin and yellow pigment via Group 2 tetraploids. The presumption is strong that the tetraploid species arose by hybridization and chromosome doubling in ancestral diploids ($2n=16$) now extinct (Fig. 1-3).

In the examples of polyploidy quoted above, the chromosome series in a given genus runs in multiples of a basic number, e.g. wheat, 7, 14, 21. It is common, however, to find more than one basic number in some genera, e.g. banana 9, 10, 11; cotton 12, 13; narcissus 7, 11; clover 7, 8, 9; vines 19, 20.

Among economic crops, *Brassica* provides an example of species with different basic numbers from which have been evolved by hybridization and polyploidy some important vegetable species. Three diploid species of *Brassica*, *B. oleracea* (cabbage group), *B. nigra* (mustard), and *B. rapa* (turnip) have haploid numbers of 8, 9, and 10 respectively, which are thought to represent a phylogenetically ascending series.

Natural hybridization between *B. rapa* (*AA*) and *B. nigra* (*BB*) has given the allopolyploid leaf mustard *B. juncea* (*AABB*). Similarly, the basic chromosome sets of *B. nigra* (*BB*) and *B. oleracea* (*CC*) have given the allopolyploid Abyssinian mustard *B. carinata* (*BBCC*); and *B. rapa* (*AA*) and *B. oleracea* (*CC*) have given the rape, *B. napus* (*AACC*). These relationships are shown in Fig. 1-4. Allopolyploids whose chromosome complements are made up of the entire somatic complements of two species are termed *amphidiploids*.

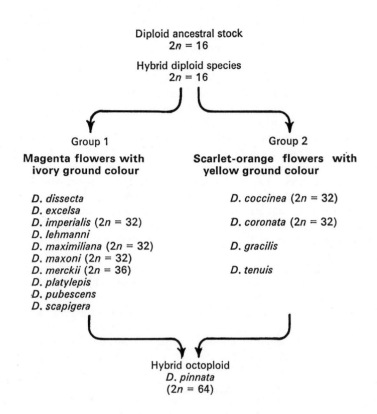

Fig. 1–3   Showing the probable origin and the constitution of the octoploid garden dahlia, *D. pinnata*. *A* and *A'*, *B* and *B'*, etc., indicate homologous chromosome sets which have differentiated sufficiently to inhibit pairing between them. Inheritance, therefore, is tetrasomic, not octosomic. Differentiation is demonstrated by the flower colour genes: *B* anthocyanin is produced in quantity, *A* anthocyanin in small amount. The *I* gene produces the flavone apigenin, the *Y* gene the chalkone butein.

In all the allopolyploids discussed in this section, chromosome doubling has occurred, or is presumed to have occurred, in the sterile hybrid between two distinct species, i.e. both parent species have directly contributed equally to the hybrid complement. This is not, however, the only way fertile allopolyploids may be derived from species crossing, as shown

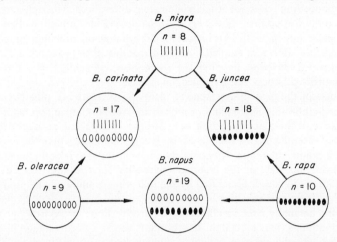

Fig. 1-4 Speciation by hybridization and polyploidy in *Brassica*. Diploid species have chromosome numbers (*n*) of 8, 9 and 10. These species have combined to form a series of *amphidiploids*, i.e. polyploids whose chromosome complement is made up of the entire somatic complements of two species. (ALLARD, R. W., 1964. *Principles of Plant Breeding*, Wiley, New York.)

by the following example. The diploid species variety, *Rubus rusticanus inermis* ($n=7$) was crossed artificially with the tetraploid species *R. thyrsiger* ($n=14$). The progeny were expected to be triploids with 21 chromosomes ($7+14$). Only four seedlings were obtained, three triploids and one tetraploid. The latter must have arisen from the union of an unreduced gamete of *R. rusticanus* with a normal gamete of *R. thyrsiger*, i.e. $14+14$ chromosomes. The three triploids were highly sterile, the tetraploid 'John Innes' blackberry highly fertile (Plate 1). In this example, chromosome doubling was *unilateral*.

To what extent unilateral doubling occurs in the wild is not known, but it has occurred naturally in a garden when an unreduced gamete of the diploid garden raspberry (*Rubus idaeus*) ($n=7$) united with a normal gamete of the octoploid *R. vitifolius* ($n=28$) to give the well-known loganberry ($n=42$).

Mendelian segregation and recombination, interspecific hybridization and polyploidy are not mutually exclusive patterns of evolution but have often been interwoven in the development of given groups of plants. Taking this into account and other factors to be mentioned later, it is not

2—P.B.

surprising that progress based on scientific breeding had to wait on extensive and intensive research. The first break-through came with the appreciation and application of the principles enunciated by Mendel, coupled with the studies initiated by Morgan and his colleagues.

Mendelism was concerned mainly with the simple inheritance of major genes controlling easily identifiable character differences. Most characters of economic importance, however, are quantitative and are controlled by many genes (see p. 25). The impact of Mendelism on breeding practice was therefore slow, and empirical methods continued to produce, and still do, improved varieties, pending the more knowledgeable application of genetic principles now fast becoming available.

Because of the great importance of plants providing food, plus the difficulties pertaining to polyploid and quantitative inheritance (pp. 25, 27), much of current plant breeding can only be carried out at large well-equipped establishments, such as State research stations, where the expert knowledge and experience of geneticists can be brought to bear on the many problems. The scale of the resources necessary for the improvement of important crops is well illustrated by the following example of the production of entirely new varieties by selection.

Lupins normally contain alkaloids which give them protection against grazing animals, hence alkaloid-free mutants are at a disadvantage in the wild, but would provide valuable fodder crops in cultivation. Von Sengbusch (VON SENGBUSCH and ZIMMERMANN, 1947) found only 6 alkaloid-free individuals out of 1,500,000 plants tested and, from the progeny of these, 10,000,000 plants were raised before one individual was found with pods that did not open and so retained the seeds (HACKBARTH and TROLL, 1956).

Breeding Aims 2

2.1 Scientific breeding

Before we proceed to discuss the techniques and methods used in controlled breeding it will be useful to outline the general problems and scope of breeding, and thus provide a background picture against which specific methods and techniques can be the better appreciated.

For almost the whole of man's existence plant improvement has been by selection, first unconscious later conscious, at least in part. Even so little as 100 years ago the details of pollination and fertilization were not generally understood in the process of reproduction and it was not until the beginning of this century that breeders came to hear of Mendel's work, and the word 'genetics' was coined by Bateson in 1906.

Thus, the practice of scientific breeding is relatively still in its infancy and for all practical purposes we may say it is but 50 years old. As plants, directly or indirectly, will for long be the main source of food, clothing, drugs, fuel and construction materials, it is obvious that the major contribution of plant breeding for the welfare of mankind lies in the future.

Much of the breeding work of the twentieth century has been, and is, exploratory in character, and it is emerging that, taking the enormous wealth of genetical variation and plasticity exhibited by the world plant population, *there should be a great expansion in breeding* as a basic necessity for a world population rapidly increasing in numbers and in its demand for a higher standard of living.

2.2 Yield

Objectives in plant breeding are manifold, varying with climate, species or crop, husbandry, economic and many other factors. A common aim of the breeder is to increase yield. The evaluation of yield depends on a variety of factors external to the plant: light (latitude); ambient and soil temperatures; wind velocity; soil texture, water capacity and fertility; the requirements of mechanization; considerations of transport; foreign competition; consumer and manufacturing (e.g. timber) preferences. To mention light and temperature alone, the influence of these on plant growth and reproduction often requires that selections should be made, at a relevant time, at two or more geographically different centres, plus the use of experimental designs (lay-out of test plots), subject to adequate statistical analysis.

In addition to factors external to the plant, yield also depends on a complex of plant characters, e.g. in cereals, grain number, grain size, number of grains per spikelet, number of spikelets per inflorescence, number of inflorescences per plant, not to mention hardiness, disease resistance and so on. Frequently, two characters may show negative correlation when selection is practised, e.g. fruit size and fruit number. Thus, while increase in yield is a major objective in many breeding programmes, its attainment requires detailed understanding of the processes involved, the appropriate methods of analysis and the necessary breeding techniques.

Half the people in the world today are undernourished and starvation is common in the undeveloped countries. Clearly, increasing yields is urgent and vital. Probably the quickest way of achieving larger yields is by improving husbandry. Even so, new varieties giving markedly increased yields could rapidly eliminate starvation if not immediately raise the general standard of nutrition.

A current example of such a possibility relates to rice. Rice is the chief staple food of 1,800 million Asian people, whose numbers are increasing at the rate of 700,000 *per week*. Rice has the highest productivity of the world's three major cereals (1964), although it is grown in a lesser area than that of wheat: wheat 279 m. metric tons, rice 267 m. metric tons and maize 225 m. metric tons. The range of yields is very great, on a national scale three to four times as much in Japan as in Laos, and much greater between individual localities. The low yields result from the use of land of poor fertility, lack of fertilizers, low yielding varieties, prejudice against improved agricultural practices and lack of capital.

The problem, therefore, of raising yields is complex and requires as a solution better husbandry, better varieties, better education and better finance. New, quick-maturing varieties are being bred which will give three crops a year instead of two, higher protein content (an urgent necessity to meet the widespread deficiency of protein foods in the Far East), adapted to different environments and resistant to disease, lodging, etc. Rice provides an example, above all others, of the way in which, on the international scale, progressive plant breeding must go hand in hand with progressive agricultural practices to realize the full potential of improved varieties.

2.3   Climate

Brief mention may be made here to an aspect of climate affecting the breeding of new varieties. The chief characteristic of climate in many parts of the world is the unpredictable variations of its components, sunshine, temperature, rainfall, etc. in a given season or year. From both the producer's and consumer's point of view the varieties of most profit are those which are reasonably tolerant of climatic variations, and crop consistently from year to year. It is false economy to grow crops which are high yielders only under the most favourable conditions. Tolerance, therefore, to short-term fluctuations in environmental conditions is a desirable aim in breeding.

2.4 Uniformity

Uniformity between individuals comprising a crop, in time of germination, rate of growth, time of flowering, fruiting and maturity, of yield, size, shape, quality, etc. is of great importance to the farmer and grower. Cultivation is easier, e.g. in the application of the pre-emergence herbicides, in thinning-out market crops where this is necessary, in steady yielding species such as the glasshouse tomato, in harvesting by maximizing the amount of flowers, seeds, fruits, vegetables ready at the same time, by reducing the amount of grading necessary as with the tree fruits, cauliflowers, lettuces, and so on.

Recently, increased production costs and shortage of labour have led to extensive mechanization of many practices in crop management. Thus, varieties of garden peas for canning are required, each of which in succession through the season produces the greatest number of pods ready for harvesting at a single time. Brussels sprouts are now being 'picked' by machine at a single harvest, hence the sprouts must be of similar size at that time (previously, those lowest on the stem matured earliest).

Mechanical shaking of fruits from apple trees and raspberry canes is coming into practice to avoid the very expensive and slow hand picking. In consequence, varieties whose fruits detach readily, but not too easily, would be an advantage (the fruits of some raspberry varieties are difficult to pick). Grading of fruits and vegetables is often statutory and is rigorously insisted upon by supermarkets and large stores, hence varieties that give produce uniform in size and quality are in increasing demand.

In some species uniformity can be achieved by using 'F_1 hybrids' between homozygous inbred lines (p. 34) and maize, onion, Brussels sprouts are examples of crops where such hybrids are playing an increasing part in meeting the demand for uniform produce. Finally, in crops that can be propagated asexually as clones (tree fruits, many bulbs and flowering plants) uniformity is ready made, so long as mutants are weeded out.

2.5 Competition and consumer demand

Mention has been made of the demand for graded produce. This is only one aspect in a rapidly changing world situation where the pressure of foreign competition may quickly lead to the dropping of one variety or crop to be replaced by another. Similarly, changes in consumer preferences may determine the demand for this or that variety, e.g. at the present time the continental apple, 'Golden Delicious', bids fair to oust some British varieties. Apples, of course, are a crop that takes years to establish.

The impact of these various demands is far greater in the case of the intensively grown horticultural crops than those whose products, e.g. flour, timber, reach the consumer indirectly; nevertheless, the time has already arrived when the rate of change in agriculture and horticulture, for

one reason or another, is likely to outstrip the rate of breeding new varieties. Under these circumstances *the study and anticipation of trends is a
new and urgent factor in the planning of breeding programmes.*

2.6 New techniques

Only brief mention will be made of the impact of these on breeding
programmes. The coming of irrigation where none existed before may
demand entirely new varieties. Short-strawed cereals have become
desirable to meet the requirements of combine harvesting. The use of CO_2
to accelerate growth and increase yield in glasshouse crops is becoming
commonplace. In the remaining decades of this century, chemical control
of plant growth will become routine, e.g. retardants which dwarf tall
flowering plants when the breeding of dwarf varieties is not possible or
economic, and therefore varieties will be needed that are susceptible to such
chemical manipulation.

Only a fraction of the aims and scope of plant breeding have been
touched upon in this chapter. Each crop presents its own special requirements with respect to breeding aims, methods, techniques and priorities.
In the future much will depend on the resource made available by en-
lightened administrators at State level, and by the knowledge and skill of
geneticists engaged in the elucidation of genetical mechanisms and the
way these should be utilized in progressive practices.

The scope, the opportunity and the challenge are potentially immense—
and immensely rewarding.

3.1 Protection

In fully controlled breeding it is essential that the flowers to be used as female and male parents are protected from contamination by 'foreign' pollen. Single flowers, or parts or all of an inflorescence, may be covered with an inverted plastic or paper bag, the mouth of which is plugged with cotton wool and secured around the stem with string or soft aluminium wire. Wire is superior to string in that it does not swell when wet and there is no need to tie it. The bags require the support of a cane or stake to avoid damage by wind (Plate 3).

A short length of drinking straw slipped over the pistil is sometimes effective in prevented unwanted pollination (e.g. in tobacco), also the drying out of the stigmatic surface. In some cases a plug of cotton wool inserted in an emasculated flower will keep off pollinating insects (e.g. in tulip), or the petals of the unopened flower may be kept tight together by a rubber band, before and after pollination (e.g. in squash).

Whole plants can be protected by the use of wood or metal frames covered with plastic mesh to keep out pollinators. Such cages should be weighted with bricks placed at the top corners to prevent the cages from being blown over by the wind (Plate 4). Large numbers of plants can be accommodated in an insect-proof glasshouse, the ventilators of which are covered with plastic mesh. A double door insect-lock is desirable. Ample ventilation openings are necessary for summer work, since the mesh considerably reduces air-change rate.

Whatever form of protection is adopted, care should be taken to see that all unwanted flowers are first removed, namely those which have shed their pollen, or whose stigmas have been exposed to foreign pollen. An inflorescence may carry more flowers than is desired to pollinate, and these must be removed, or on opening they will shed their pollen and spoil the desired pollinations.

3.2 Isolation

Mass self- or cross-pollination may be achieved by growing a block of plants sufficiently remote from any likely pollinators, cultivated or wild, as to ensure total or high control of the desired selfs or crosses. The safe distance for spatial isolation depends in the first place on the reproductive system of the plant, on the kind of pollinators and their habits, and on population size and density.

Investigation has shown that whether a species is insect- or wind-pollinated, intercrossing falls off with distance in a regular fashion (Fig. 3–1). Where the proportion of contamination (F) by outcrossing is small,

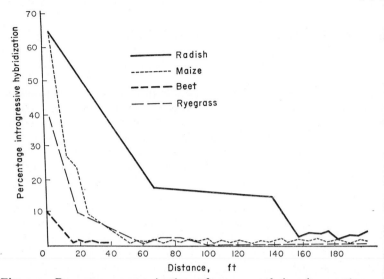

Fig. 3–1 Percentage contamination of one population by another at different isolating distances. Test progeny uniformly homozygous for a recessive allele were placed at different distances from contaminating populations of homozygous, dominant genotypes and the percentage of contamination scored in the progeny of the test population. (WILLIAMS, W., 1964. *Genetic Principles and Plant Breeding*, Blackwell, Oxford.)

contamination in both insect- and wind-pollinated species is given by the formula, $F = y/De^{KD}$, where D = distance, y = contamination at zero distance, e is a mathematical constant, i.e. the exponential function, and K the rate of decrease of contamination with distance. This formula allows the prediction of contamination at any distance, provided the amount of contamination at any two points is known. Spatial isolation is often used in the bulking up of seed stocks.

3.3   Emasculation

Most flowering plants bear hermaphrodite flowers, and it is necessary to remove the anthers from those flowers selected as females. This is usually done with fine pointed forceps or scissors (Fig. 3–2). Both forceps and fingers should be sterilized in methylated spirits before starting emasculation, and as often as different varieties are handled.

Flowers should be emasculated as late as is safe before the pollen is shed. Often petals, and perhaps sepals, must be removed to permit emasculation. Occasionally, this may result in unacceptable damage, e.g. in raspberry the petals may be removed, but not the protective sepals, as the styles then dry up and are rendered useless. Fuller details on the subject and on the keeping of records are given in LAWRENCE (1957) and ALLARD (1964).

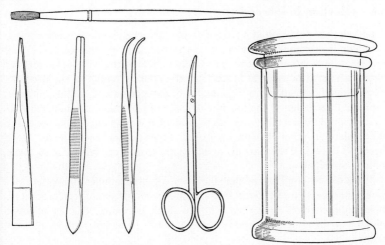

Fig. 3–2 The plant breeder's tools—forceps, brush, scissors, and a wide-mouthed jar for methylated spirits. (LAWRENCE, W. J. C., 1957. *Practical Plant Breeding*, Allen & Unwin, London.)

3.4 Self-pollination

In a number of species, the reproductive systems exhibit devices for ensuring self-pollination, e.g. cleistogamy in lettuce and violets, juxtaposition of anthers and stigmas, anthers which dehisce before extrusion from the flower (wheat), or the stamens form a tight cone around the stigmas (tomato). In self-pollinated plants such as these it is rarely necessary to protect the flowers from pollinating agents. In other self-pollinated plants, e.g. those where the anthers shed their pollen directly on to the stigma, there is always the risk of cross-pollination, hence protection and artificial self-pollination are a necessity in controlled breeding. Various techniques used in self-pollination are mentioned in the section on cross-pollination which follows. Examples of self-pollinated species are barley, citrus, flax, oats, parsnip, pea, peach, peanut, sweet pea, tobacco, wheat.

3.5 Cross-pollination

The majority of plants are outbreeders to a great extent. Some exhibit specific devices that encourage outbreeding, e.g. protandry (carrot, parsnip), protogyny (walnut), monoecy plus protandry (maize), dioecy (hops, asparagus, date palm) self-incompatibility (many species, e.g. cherry, red and white clover, almond, cabbage, radish). Obviously, where there is a suspicion that cross-pollination may occur, however little, then protection of male and female flowers is a prerequisite to fully controlled breeding. For details see LAWRENCE, 1957.

3.6 Selection in inbreeders

Clearly, the selection of a parent or parents in a breeding programme is of major importance in determining the potential of the progeny. What must not be overlooked, however, is that this potential will not be realized unless accurate assessment is made of the relative merits of the progeny. Such judgement calls for a high degree of perception and skill based on an intimate knowledge of the morphological and physiological characteristics of the crop. Often the breeder has to deal with large numbers of plants and then selection must depend on quick, expert, visual assessment. Good breeders develop a flair for this exacting work: they are artists, so to speak, as well as scientists.

It should be emphasized that the breeding programme appropriate to a given species is largely determined by its breeding system, e.g. whether it is an inbreeder or an outbreeder. We shall, therefore, discuss selection under two headings: in self-pollinated crops (3.6), in cross-pollinated crops (3.7).

3.6.1 Single plant selection

The genetic effect of continued self-fertilization in self-pollinated species or cultivars is to sort out and reveal the dominant and recessive genes. As Mendel showed, selfing the heterozygote, Aa, gives in the next generation (F_1) the three genotypes, AA, Aa, and aa in the ratio $1:2:1$. On selfing each of these genotypes, AA and aa breed true, while Aa gives

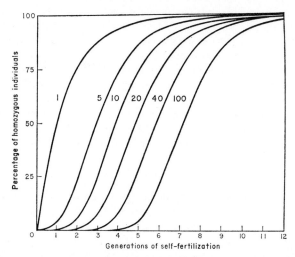

Fig. 3–3 Percentage of homozygous individuals after various generations of self-fertilization, when the number of independently inherited heterozygous gene-pairs is 1, 5, 10, 20 and 100. (ALLARD, R. W., 1964. *Principles of Plant Breeding*, Wiley, New York.)

1AA:2Aa:1aa as before. Thus, heterozygosity is reduced by one half in each generation and after 6 or 7 generations of selfing, the population will consist almost entirely of equal numbers of the two homozygotes AA and aa, i.e. homozygosity is rapidly achieved by selfing (Fig. 3–3).

When the many thousands of genes in an inbreeding species are taken into account, it is evident that the population will consist of many different homozygous genotypes, or *lines*: genetic differentiation is *between* varieties (lines) and not *within* them. This applies to polygenes (minor genes, see p. 26) as well as to major genes; to quantitative as well as to qualitative characters. Selection of characters, therefore, by continued selfing results in homozygous progenies, i.e. in *pure lines*. When a new or desired character is selected in this way, it is said to be 'fixed', and the character breeds true. The amount of variation (lines) possible among a group of inbreds is indicated by the fact that a single parent possessing only 10 heterozygous loci (a locus is the position occupied by a gene in a chromosome) can give 1,024 different homozygous genotypes (lines).

The single plant selection method with inbreeding crops is to select a *large* number of superior plants from the genetically variable original population and to raise self progenies of each of these, preferably in different environments. The selection of superior lines is made each year, until the breeder can no longer decide between lines from observation alone. Resort must then be made to replicated trials of the remaining lines, especially comparing them with established commercial varieties, and for more than one season.

The foregoing procedure will demand the most care and time where the characters for which selection is being made are inherited quantitatively. The procedure is shortened when it is desired to fix, say, one conspicuous mutant character controlled by a major gene, the object then being to select for general plant performance among the progenies carrying the mutant character.

It should be noted that the consequences of artificial selection in a self-pollinating species or crop, depend on the initial genetical variability in the crop at the beginning of the breeding programme. Nothing, except occasional mutation, can be added to the gene pool.

3.6.2   Mass selection

The simplest form of mass selection is that in which inferior or off-type plants in a crop are destroyed (culled, or rogued-out) before flowering begins. Time and labour can be saved when there is correlation between juvenile and adult characters, e.g. yellow- and white-flowered varieties of dahlia and snapdragon are without anthocyanin in stems and leaves in the seedling stage and rogueing-out of unwanted red-flowered variants (or vice versa) is possible very early. Mass selection often has the advantage of permitting the retention of the best features of an original variety with respect to general adaptation (p. 51) and avoiding the extensive testing required in pure line breeding.

In the mass selection of self-pollinated crops it is important to maintain an adequate number of lines, otherwise there is danger of impairing the essential features and adaptation of the variety. Selection, therefore, must be rigorous enough to eliminate obvious off-types, but not so rigorous as to reduce the number of lines and thus narrow the genetic base. Indeed, the presence of numerous closely related pure lines ('multi-lines') in mass selected varieties would be expected to impart a useful measure of adaptability in the response of the crop to minor changes in the environment.

3.6.3 Pedigree breeding

The most widely used method of breeding in self-pollinated plants, by both amateur and professional breeders, is the pedigree method (cf. animal breeding). Superior types are selected in successive segregating generations as in pure line breeding, and a record, the pedigree, kept of parent-progeny relationships and performance.

Pedigree breeding starts with the crossing of two varieties which complement one another with respect to one or more desirable characters, and in the F_2 generation single plant selection is made of those individuals which the breeder judges will produce the best progeny. In F_3 and F_4 generations selection is practised for the best plants in the best families. By the F_5 or F_6 generations, most selections will be homozygous, i.e. selection will no longer be effective within the families, only between them.

If the two original parents do not combine all the desired characters, then a third parent may be included by crossing it with an F_1 individual, or a fourth parent introduced by crossing two F_1's.

3.6.4 Bulk population breeding

In the commercial breeding of grain and bean crops, the bulk population method has two advantages: it avoids the labour entailed in pure line and pedigree breeding, and depends for its success on the operation of natural selection.

The method entails the making of an F_1 and the raising of an F_2 comprising a large number of individuals. The crop is harvested in bulk and the seeds are sown in similar quantity the following year, and repeated in as many years as the breeder thinks fit. Natural selection will reduce or eliminate those types which have a poor survival value. To a minor extent artificial selection can augment the process, by the rogueing-out of obviously inferior types or off-types. Artificial selection can also be used to select, inexpensively, for earliness, by harvesting when part of the crop has matured. By growing the seed from the F_2 and subsequent generations in different localities, natural selection will result in those varieties best adapted to a given locality becoming predominant.

3.6.5   Backcross breeding

Not infrequently it is desired to improve a variety having many good qualities, but lacking in one or more, by transferring a desirable character from another variety otherwise not commendable. In general terms, this is accomplished by making a series of backcrosses of the inferior (donor) variety to the superior one (recurrent parent), selecting for the desired character at each generation. After a sufficient number of backcrosses the progeny will be heterozygous for the alleles it is desired to transfer, but homozygous for all others. Selfing the last backcross generation, coupled with selection, will then give some progeny homozygous for the genes being transferred, and identical with the superior variety in all other respects. It must be possible, of course, to identify the character being transferred in the successive backcrosses, even though it may be temporarily diminished in expression.

3.7   Selection in outbreeders

3.7.1   Single plant selection

The requirements for, and results of, selection in cross-pollinated species are very different from those with self-pollinated ones. This ensues from the nature of the outbreeding population, which has been defined as a community of sexual and cross-fertilizing organisms *which share a common gene pool*. In such a population, single plant selection and segregation often cause the progeny to deviate from the parental type with adverse effects on vigour, fertility and productivity, unlike the case with single plant selection in naturally self-pollinated species. The problem, therefore, is to devise breeding programmes that permit selection towards a high frequency of favourable gene combinations without impoverishing vigour, fertility and productivity as a whole.

Each cross-pollinated plant is heterozygous for a great many genes and continued inbreeding often results in loss of vigour, styled *inbreeding depression*, sometimes to the point of extinction of some lines, due to the segregation of recessive genes in homozygous condition whose deleterious effects are normally suppressed by their dominant alleles (*dominance hypothesis*). An alternative hypothesis is that heterozygous combinations of alleles at a given locus, e.g. *Aa*, are superior to either of the homozygous ones, *AA* and *aa*, and vigour increases in proportion to the total heterozygosity in the individual (*overdominance hypothesis*).

Theories aside, it is essential with cross-pollinated plants that the breeding programme maintains heterozygosity. Consequently, single plant selection can only be practised in a modified form, e.g. it may be possible to inbreed for a while, selecting the best phenotypes, but then these must be intercrossed to re-establish an adequate amount of heterozygosity. The position is further complicated when the species is self-incompatible (p. 30) since, with rare exceptions, single plant selection is then prohibited.

In this case, a number of plants must be selected which together combine the desired attributes, and then intercrossed. Only experience with a given outbreeding species can determine the amount of inbreeding which is practicable in that species.

3.7.2.   Mass selection

Over the centuries, mass selection in cross-pollinated plants has undoubtedly been effective in crop improvement (cf. sugar-beet, p. 5), especially in the adaptation of varieties to new localities. It has the advantages of requiring only one cycle of selection per generation and of minimal labour requirements together with simplicity of operation. Mass selection has proved useful when the character to be improved upon can be readily seen or measured and recently, contrary to earlier opinion, it has also proved to be effective in increasing yield, a quantitative character, in maize. Thus, four generations of selection increased yield by 22 per cent in one case, and 33 per cent in another after three cycles of selection. These are orders of magnitude similar to those obtained by the substitution of hybrids (p. 33) for open-pollinated varieties (SPRAGUE, 1966).

Some limitations of mass selections in cross-pollinated crops can be overcome, up to a point, by *progeny selection*. Progeny testing is an invaluable technique for distinguishing between parents whose apparent superiority is due to the environment and not the genotype, i.e. between non-heritable and heritable variation.

Among the various mating systems that can be used to obtain progenies for testing, *line breeding* is an important method with cross-pollinated plants. In brief, this consists in mass selection for several generations, followed by saving seed from the most superior plants and sowing this mixed, or *composited*, seed in an isolated plot where the plants mate at random. Line breeding avoids the ill effects of excessive inbreeding, so long as an adequate number of not too closely related lines is included in the composited seed. In a number of crops, this technique is of great importance in maintaining a high level of resistance to pathogens.

3.7.3.   Recurrent selection

A method that combines the merits of progeny testing and line breeding, together with a measure of control over pollination, is that of *recurrent selection*. In its simplest form, this consists in selecting superior plants from a heterozygous population and propagating them by selfing. The selfed progenies are then intercrossed in all combinations to give material for further cycles of selection and intercrossing.

An improved form of this procedure is to start with two genetically unrelated heterozygous source populations, *A* and *B*. A number of plants from source *A* showing desirable characteristics are self-pollinated and also crossed with a small random sample of plants from source *B*. Similarly, plants from source *B* are selfed and also crossed with a random sample from source *A*. The selfed seed is kept in reserve. In the second year, a replicated trial plot is set up to compare the performance of the *crossed* progenies derived from the *A* source. A similar plot is set up for the crossed progenies from the *B* source. The *parentage* of the superior plants in each plot is noted. In the third year, selfed seed obtained in the first year from the superior source *A parents* is sown in a block and all possible inter-crosses are made within this block. The same procedure is carried out with the superior *parents* from the *B* source. In the fourth year, the seed from the intercrosses within each block is sown to give two new sources, A^1 and B^1. This completes the cycle of operations and the same procedure as before is followed in further cycles. Commercial seed is produced from crosses between the A^1 and B^1 source groups. This method of *reciprocal recurrent selection* consists in essence of using for line breeding only those parents shown to be superior by progeny tests.

3.7.4 Backcross breeding

The method of backcross breeding in cross-pollinated crops differs from that used for self-pollinated crops only in that a number of plants must be used as recurrent parents, instead of one plant. This is necessary to ensure that the sample of gametes carried by the recurrent parents represents the gene frequency characteristic of that variety. Backcross breeding in cross-pollinated crops has been widely used to transfer genes for disease resis-tance into established, but susceptible, varieties (see ALLARD, 1964).

3.8 Quantitative inheritance

Mendel and the early plant breeders of this century dealt with readily identifiable character differences controlled by major genes, that is with discontinuous variation and qualitatively different characters; dwarf and tall peas, red- and yellow-fruited tomatoes, coloured and white flowers.

From the earliest days, however, it was obvious that many characters intergraded in expression and in inheritance: fruit size (Fig. 3–4) and colour, doubleness of flowers (Plate 5), plant stature, cold and drought

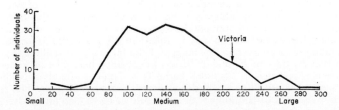

Fig. 3–4 Showing range of variation in fruit size among 211 seedlings derived from selfing the plum, Victoria. (CRANE, M. B. and LAWRENCE, W. J. C., 1952. *Genetics of Garden Plants*, Macmillan, London.)

tolerance, time of maturity, and so on. Inheritance was continuous and quantitative, as well as discontinuous and qualitative in character.

As Mendelian studies progressed to the discovery of multiple factors, and modifying genes, it became clear that quantitative and qualitative categories were not of nature's making, but man's. Mendelian ratios became difficult to identify when several genes were involved because of overlapping of phenotypes; overlapping which might be the more confused by non-heritable variation due to environmental effects.

It is, nevertheless, convenient to refer to 'major' genes controlling characters clearly exhibiting Mendelian inheritance and dominance, but to these we must now add a second arbitrary category, namely, *polygenes* whose effects are too small to be recognized individually and which may show no dominance but have similar or supplementary effects.

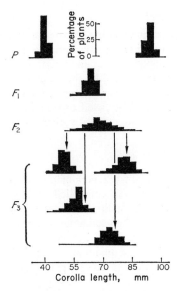

Fig. 3–5 Frequency distribution of corolla length in families of *Nicotiana longiflora*, showing segregation for the polygenes controlling the differences between the two parent strains (P) in this character. Variation in P and F_1 is wholly non-heritable. To this is added in F_2 the heritable component depending on segregation. The spread of the frequency distribution is thus higher in F_2 than in F_1 and P. In F_3 the average heritable component of variation is half that in F_2, but it varies from family to family according to the number of genes for which the F_2 parent was heterozygous. The spread of the frequency distribution is thus variable in F_3 though always lying between that of F_2 on the one hand and of F_1 and P on the other. Note that the F_3 means are correlated with the values of the F_2 individuals from which they came. (DARLINGTON, C. D. and MATHER, K., 1950. *The Elements of Genetics*, Allen & Unwin, London.)

(*above*)

Plate 1 Sister seedlings from the cross *Rubus rusticanus inermis* ($2n = 14$) x *R. thyrsiger* ($2n = 28$). RT3 is a triploid ($2n = 21$) and sterile. RT4 is a tetraploid ($2n = 28$) and fertile. (CRANE, M. B. and LAWRENCE, W. J. C., 1952. *Genetics of Garden Plants*, Macmillan, London.)

(*left*)

Plate 2 Bud pollination in radish (*Raphinus sativus*). The inflorescences to the left were pollinated when the flowers opened. The inflorescence to the right was pollinated when the flowers were in the bud stage. Note difference in seed set. (KAKIZAKI, Y. and KASAI, T., 1933. *J. Heredity*, **24**, 359, American Genetic Association and Allen & Unwin, London.)

Plate 3 A method of protecting flowers from insect visitors by the use of transparent plastic bags. *Dahlia.* (LAWRENCE, W. J. C., 1957. *Practical Plant Breeding*, Allen & Unwin, London.)

Plate 4 Muslin (or plastic) covered cages used for protecting plants in the open ground from insect visitors. (LAWRENCE, W. J. C. 1957. *Practical Plant Breeding*, Allen & Unwin, London.)

Polygenic characters can be specified accurately only in terms of metrics such as weight, volume, length, proportions, time, etc., and are called metrical characters. Polygenic characters are inherited in the same way as those controlled by major genes, but require a different method of analysis, namely biometrical analysis, to distinguish, for example, between heritable and non-heritable effects. Biometrical analysis is concerned with numerical quantities, or parameters, such as means, variances and co-variances and their genetic content, a subject which, though in detail is outside the scope of this book, is of great importance to the plant breeder who, more often than not, finds himself dealing with quantitative characters (Fig. 3–5).

It may be briefly noted here that polygenic systems have specific properties, one of which is the storage of variability by the action of plus or minus alleles of different genes balancing one another in the genotype. Recombination frees the latent variability by upsetting the balance, so that polygenes working in the same direction are brought together and the action of selection thereby facilitated.

3.9 Polyploid inheritance

Two main types of polyploids are polyploid varieties or *autopolyploids* and polyploid hybrids or *allopolyploids*. Each of these may be further sub-divided into *euploids* with a whole number multiple of the basic chromosome number (x) of the group (e.g. tetraploid, hexaploid, octoploid), and *aneuploids* whose chromosome complement is not a whole number multiple of the basic number (e.g. $4x + 1$, $6x - 1$) (Table 2).

Table 2 Some polyploid types. In this table, x = the basic chromosome number = 3 non-homologous chromosomes, A, B and C. The nullisomic is deficient in both C chromosomes; the monosomic is deficient in one C chromosome. The trisomic has 1 extra A chromosome, the tetrasomic 2 extra A chromosomes. The monoploid and diploid formulae are included for comparison with the other types.

Type	Formula	Somatic chromosome complement
Euploids		
monoploid	x	ABC
diploid	$2x$	ABC, ABC
triploid	$3x$	ABC, ABC, ABC
autotetraploid	$4x$	ABC, ABC, ABC, ABC
allotetraploid	$2x + 2x'$	ABC, ABC, ABC', ABC'
Aneuploids		
nullisomic	$2x - 2$	AB, AB
monosomic	$2x - 1$	ABC, AB
trisomic	$2x + 1$	ABC, ABC, A
tetrasomic	$2x + 2$	ABC, ABC, A, A

Euploids with an odd number of chromosome sets (triploids, penta-ploids, septaploids) are characterized by a high degree of sterility owing to irregular meioses and unbalanced gametes; hence it is rarely practical to attempt a systematic study of their inheritance. Nevertheless, some odd-numbered euploids are of great importance in crops which can be pro-pagated asexually. For example, some apple and pear varieties, commercial bananas and numerous decorative plants are triploids showing greater vigour and size than their comparable diploid types. They also comprise highly uniform clones, an advantage in such crops.

Turning to euploids with an even number of chromosome sets, we need only consider inheritance in the simplest of them, the *autotetraploid* and the *allotetraploid*. Neither need we take into account the various complexities of chromosome pairing at meiosis, only the consequences in the production of the number of kinds of gametes.

The autotetraploid usually arises from the union of two unreduced gametes in a diploid, or from somatic chromosome doubling. In either case there are four identical sets of chromosomes $CCCC$, instead of two CC, and correspondingly the gametes, aberrant pairing apart, contain two sets of chromosomes instead of one.

With random chromosome pairing, the consequences of Mendelian inheritance are somewhat different in the autotetraploid compared with the diploid. The familiar $3:1$ and $1:1$ ratios will occur in the appropriate progenies, also others not found in diploids. For instance, the diploid heterozygote Aa produces equal numbers of A and a gametes and on selfing, the resultant progeny comprise three genotypes in the ratio $1AA:2Aa:1aa$, i.e. the product of the gametic series $(A+a)(A+a)$ or $(A+a)^2$. In the autotetraploid, selfing the comparable heterozygote $AAaa$ gives three kinds of gametes, AA, Aa and aa in the ratio $1:4:1$. Hence the genotypes given by $(AA+4Aa+aa)^2$ equal $1AAAA: 8AAAa: 18AAaa: 8Aaaa: 1aaaa$, i.e. a phenotypic ratio of $35A:1a$. Other phenotypic ratios are $5:1$ from $AAaa \times aaaa$ and $11:1$ from $AAaa \times Aaaa$. Comparing inheritance in diploid and autotetraploid varieties, more plants must be raised to extract the bottom recessive (or top dominant) genotypes from selfing (or backcrossing) the middle heterozygotes, namely, 9 times as many with one pair of alleles, 81 (9^2) times with two pairs and 729 (9^3) times with three pairs of alleles. In the autotetraploid, when dominance is complete, $AAAA$, $AAAa$, $AAaa$ and $Aaaa$ genotypes will be indistinguishable phenotypically. When dominance is incomplete, it may be possible to recognize each of these types, so long as non-heritable variation does not obscure their differences.

When duplicate, complementary, and multiple genes along with chromatid segregation and linkage are taken into account, it will be clear that the plant breeder is confronted with some formidable problems of analysis in autopolyploids, and that character expression will often be quantitative. Many domesticated species are autotetraploids, since they are often larger than their diploid counterparts, and have been selected for this reason.

The other kind of simple polyploid is the allotetraploid, arising from the crossing of different diploid species. Usually such diploid hybrids are more or less sterile: their chromosome sets are too different structurally to pair normally at meiosis. We may represent the two sets in the hybrid as M and M^1. Should somatic doubling of the chromosomes occur in the sterile hybrid, or two unreduced gametes unite to give an individual with four sets MMM^1M^1, instead of two, then fertility is restored as the M set can now pair with M, and the set M^1 with M^1. In consequence, the allo-tetraploid tends to behave as a functional diploid in inheritance.

A notable example of this is the octoploid (double autotetraploid) garden dahlia mentioned earlier (p. 9). In a high polyploid of this nature it is essential for its survival that meiotic regularity is of a high order. In dahlia, only one chiasma is formed in each arm of the chromosomes and all chiasmata are terminalized. This makes for easy chromosome disjunction at the first metaphase of meiosis and the paired chromosomes disjoin with great regularity. In consequence, the garden dahlia is highly fertile and the tetrasomic ratios correspondingly good. Incidentally, dahlia is also highly heterozygous, since it is self-incompatible.

The garden dahlia exemplifies the manner in which natural selection has operated to remove any tendency to meiotic irregularity, and achieve all-round balance in a highly polyploid and highly efficient species. It will be remembered that whereas the autotetraploid is similar to its diploid progenitor in its gene content, the allotetraploid combines the gene content of two differentiated species, hence its potential capacity for variation is greater.

Dahlia also illustrates a phenomenon common to high polyploids. Not once during the 170 years since its introduction to Europe, and in spite of the great amount of work by breeders, has a mutation abruptly appeared (cf. the diploid sweet pea). New characters first occur with a minimum of expression, e.g. a few extra petals in single flowers, and selection is neces-sary, generation after generation, until the mutant double-flowered form is well-expressed and the genotype relatively homozygous.

For simplicity's sake, reference has been made to inheritance in 'true', or perfect, auto- and allotetraploids. The distinction between them, however, is somewhat artificial, e.g. hybridization of two species whose genomes have differentiated in evolution will, on chromosome doubling, give an allopolyploid in which pairing may be selective rather than random (see WILLIAMS, 1964, and ALLARD, 1964).

Breeding Methods: Special 4

4.1 Incompatibility

As pointed out (p. 19) many wild species are outbreeders. One of the most efficient mechanisms for securing outbreeding is that in which the plant's own pollen, though perfectly viable, does not produce pollen tubes capable of growing down the style to effect fertilization, i.e. the pollen is *incompatible*. It has been estimated that *self-incompatibility* is present in more than 3,000 species of flowering plants.

Several incompatibility systems are known (Allard 1964) but only one, the most important, will be discussed here. In this system, incompatibility is governed by a single gene, S, which has a number of alternative forms, called *multiple alleles*, at its locus, namely S_1, S_2, . . . S_n. In brief, pollen cannot function in a style carrying an incompatibility gene common to both pollen and stylar tissue. Consequently almost all plants in which this system is found are heterozygous at the S locus, e.g. S_1S_2, S_1S_3, but not S_1S_1, S_2S_2. The result of this is that three types of pollination may be identified; $S_1S_2 \times S_1S_2$ in which all pollen is incompatible, $S_1S_2 \times S_1S_3$ in which half the pollen functions, and $S_1S_2 \times S_3S_4$ in which all pollen is compatible. The progeny of $S_1S_2 \times S_1S_3$ and its reciprocal will comprise both parental genotypes and one new one, S_2S_3, while the progeny of $S_1S_2 \times S_3S_4$ will all be different from the parental genotypes (Fig. 4–1).

An accompaniment of self-incompatibility in certain matings is *cross-*

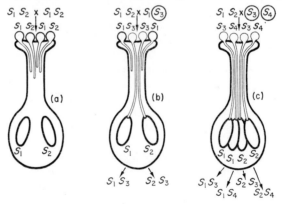

Fig. 4–1 Diagram of pollen-tube incompatible and compatible pollinations. (CRANE, M. B. and LAWRENCE, W. J. C., 1952. *Genetics of Garden Plants*, Macmillan, London.)

incompatibility. This circumstance is of great importance in some crops (CRANE and LAWRENCE, 1952). In the diploid sweet cherry, the varieties fall into groups in any one of which all varieties have the same constitution for incompatibility (Fig. 4–2). Since all sweet cherries are self-incompatible, the fruit grower must know which of them are *cross-compatible* to secure

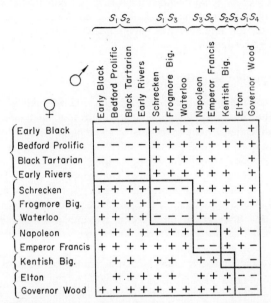

♀ \ ♂	Early Black	Bedford Prolific	Black Tartarian	Early Rivers	Schrecken	Frogmore Big.	Waterloo	Napoleon	Emperor Francis	Kentish Big.	Elton	Governor Wood
	S_1S_2				S_1S_3			S_3S_5		S_2S_3	S_1S_4	
Early Black	−	−	−	−	+	+	+	+	+	+		+
Bedford Prolific	−	−	−	−	+	+	+	+	+	+	+	+
Black Tartarian	−	−	−	−	+	+	+	+	+			+
Early Rivers	−	−	−	−	+	+	+	+	+	+		+
Schrecken	+	+	+	+	−	−	−	+	+	+	+	+
Frogmore Big.	+	+	+	+	−	−	−	+	+	+	+	+
Waterloo	+	+	+	+	−	−	−	+	+	+		
Napoleon	+	+	+	+	+	+	+	−	−	+	+	−
Emperor Francis	+	+	+	+	+	+	+	−	−	+	+	−
Kentish Big.		+	+		+	+		+	+	−		−
Elton		+	+	+	+	+		+	+	+	−	−
Governor Wood	+	+	+	+	+	+	+	+	+	+	−	−

Fig. 4–2 The compatibility relations of 12 varieties of sweet cherry, falling into 5 incompatibility groups. + indicates successful pollination, − unsuccessful pollination, and a blank that the pollination has not been tried. The compositions of the varieties relative to the S gene are shown above. (DARLINGTON, C. D. and MATHER, K., 1950. *The Elements of Genetics*, Allen & Unwin, London.)

adequate cross-pollination in the orchard, i.e. a Group 1 variety must be planted with, say, a Group 2 pollinator. The grower must also select varieties which flower at the same time.

Self-incompatibility may not be absolute under all conditions. Bud pollinations may allow slow-growing pollen tubes enough time to reach the ovules (Plate 2). Similarly, the application to the base of the style of an aqueous solution of beta-naphthoxy-acetic acid at 40 p.p.m. may delay abcission of the carpel, again giving more time for pollen tube growth. Or the style may be cut off, the cut stump pollinated and 1 per cent alpha-naphthyl-acetamide in lanolin used to hinder abscission of the flowers and encourage the development of the seed vessel. When the site of the incompatibility reaction is confined to the region of the stigmatic surface, removal of this may permit an adequate rate of pollen tube growth. End of season pollination, and pollination at low temperature, are among other practices that sometimes overcome incompatibility.

The functioning of the S incompatibility system outlined above, refers to diploids in which only one S allele is carried in each pollen grain. In auto- and allopolyploids, two or more S alleles are carried in each pollen grain and interaction may occur between them leading to the partial or complete breakdown of incompatibility. Breakdown is most likely in newly created polyploids; natural selection will have tended to eliminate interaction in long established polyploids.

4.2   Interspecific sterility

Sterility commonly occurs in interspecific hybrids and is due to un-balance of the chromosome complement, resulting in such effects as abor-tion or modification of entire flowers or their parts, or disruption in the development of pollen, embryo sac, embryo or endosperm. Sterility of this kind can also ensue from excessive inbreeding.

Plants of outstanding value have frequently resulted from inter-specific crosses. Often such hybrids are sterile, but this may be no draw-back where asexual propagation is possible, e.g. in shrubs, bulbs. More often than not, attempts at interspecific crossing meet failure to secure seed, but initial failure should not be accepted too readily, as repeated efforts may eventually yield a few seeds.

When two parent species have different chromosome numbers, it not infrequently happens that while $A \times B$ fails, $B \times A$ gives some seed. Per-sistence in pollinating also applies to apparently sterile species hybrids. Failure of seed development in interspecific hybrids may sometimes be due to unfavourable interaction between embryo and endosperm. Excis-ion of the embryos which are then cultured on solid nutrient media until they are large enough to transfer to soil is a method that has been used successfully with some species hybrids. Spontaneous chromosome doubling is by no means uncommon in species hybrids, and the chance of finding doubled forms increases with the number of plants of the hybrid available. The use of mutagenic agents may be a more positive approach to securing fertile forms from sterile hybrids (see p. 36).

4.3   Male sterility

Some forms of male sterility are of considerable advantage or potential, both to the grower and the breeder. From time to time male-sterile individuals have appeared in a number of genera (carrot, cucumber, sugar-beet, tomato, barley, flax, sorghum, cocksfoot) as the result of mutation of a gene, or genes, governing the formation of pollen. Though such mutants would not survive in nature they can provide the breeder with valuable material for the mass production of F_1 hybrid seed (p. 36).

In many crop plants, male sterility is governed by a single gene and this condition can be maintained by crossing the recessive male-steriles, ss, with heterozygous fertile plants, Ss, giving half male-sterile and half fertile progeny. In utilizing male sterility for hybrid seed production, the male-sterile line is planted in alternate rows with the desired male parent, and the fertile plants in the male-sterile rows are rogued as soon as they can be identified.

Another type of male sterility depends on cytoplasmic factors. In the simplest case, the male-sterile line will produce seed if pollinators are present, but because the cytoplasm is inherited only maternally, the resulting progeny are all male-sterile.

A third type of male sterility depends on the interaction of gene and cytoplasm. In onion, there occurs a pair of genes, R and r, and two kinds of cytoplasm, fertile (F) and sterile (S). The genes are inherited through both parents in the usual Mendelian way, but the cytoplasm is inherited through the female parent only, i.e. $(F) \times (S)$ gives (F), and $(S) \times (F)$ gives (S). All plants with (F) cytoplasm produce viable pollen, hence (F) RR, (F) Rr and (F) rr are male-fertile. But plants with (S) cytoplasm are male-fertile or male-sterile according to whether they carry the dominant gene R or the recessive gene r. Thus, (S) RR and (S) Rr are male-fertile, whereas (S) rr is male-sterile, i.e. interaction between the r gene and (S) cytoplasm results in bad pollen and male sterility. Or to put it the other way, the R gene restores the pollen-producing ability of the (S) cytoplasm and is referred to as the *restorer* gene. Male sterility of this kind is also found in cocksfoot grass, flax and sugar-beet.

(F) rr forms exist in most commercial onion varieties, so by using the original (S) rr male-sterile form as female parent and crossing it with different male-fertile varieties of the constitution (F) rr, male-sterile offspring can be produced carrying the genes which govern the desirable characters of the commercial varieties. By continued backcrossing, male-sterile lines are obtained practically identical with the established commercial varieties, thus facilitating the large-scale production of F_1 hybrid varieties (see ALLARD, 1964).

4.4 Heterosis

Mention was made on p. 23 of the loss of vigour which often follows the inbreeding of naturally outbreeding varieties. The opposite of this, the increase in vigour which may occur in the F_1 from crossing two different varieties, is generally described as *heterosis*. Strictly, heterosis may be negative or positive; the F_1 may be less vigorous than the parents, or more vigorous. The term, however, is conventionally reserved for describing positive heterosis, or hybrid vigour.

The effects of heterosis may be various; earlier germination, increased rate of growth, increased size of leaves, greater stature, greater uniformity, earlier flowering, increased yield, the last being of greatest interest and importance. The maximum expression of heterosis is in the F_1, succeeding generations declining in vigour to the parental level. As we shall see, heterosis cannot, as a rule, be predicted from parental appearance or performance. Usually, however, it is found in the F_1's of parents neither distantly nor closely related, and in outbreeding varieties.

The exploitation of heterosis in commercial crops depends in the first place on the development of *inbred lines*, as follows. In outbreeding species, continued inbreeding, say by selfing, results in the segregation of a percentage of abnormal forms; deficient root systems, dwarfness, chlorophyll deficiency in various degrees, abnormal flowers, defective seeds, infertility. These deviants are the expression in homozygous form of deleterious recessive genes, whose expression is suppressed in the parents by their dominant alleles. Inbreeding, therefore, purges the line of these undesirable genes.

By selfing each generation and selecting normal and productive forms possessing the most desirable characters, inbred lines can be established which, though different from one another, are highly uniform themselves, i.e. are homozygous. One important consequence, therefore, of the development of inbred lines is that when any two lines are crossed together, the F_1's also are highly uniform, and this uniformity, as we saw earlier (p. 15), is a valuable characteristic in commercial varieties.

Once a number of inbred lines have been established the remaining problem is to discover which of them when crossed, give the greatest vigour in F_1 along with, of course, a combination of the most desirable characters. A great deal of research has been directed to the discovery of methods that would eliminate the immensely wasteful procedure, in time, labour, material and space, of crossing lines in all combinations to ascertain which ones give the most vigorous F_1's in terms of yield. By far the greatest use of hybrid varieties has been in maize and some breeding methods used with this species will serve to illustrate the approach to the prediction of heterosis.

Maize is a monoecious species in which the male flowers or *tassels* are borne at the top of the main stems, while the female flowers or *ears* are borne along the sides. Pollination in maize is by wind which blows pollen from the tassels on to the long styles, called *silks*, which protrude from the tops of the ears. But because the tassels on a given plant shed their pollen 2–3 days before the silks are receptive, these are usually pollinated by pollen from other plants, i.e. maize is naturally cross-pollinated.

The production of hybrid maize comprises three stages: (a) the selection of superior plants in open-pollinated populations, (b) selfing these plants for several generations to produce homozygous inbred lines, and (c) crossing selected lines. Hybrids may be produced by crossing two inbred lines $A \times B$, or by making a three-way cross between an inbred line as female parent and another variety as male $(A \times B)C$, or, usually, by crossing two single-cross inbred lines $(A \times B)$ $(C \times D)$ (Plate 6).

Numerous investigations have revealed that although a small measure of success with hybrid maize can be attributed to the selection practised in the production of inbred lines, the greater part of the increased yield from double-crosses is due to the use of inbred lines with a good *general combining ability*. The identification of such lines is of paramount importance and is achieved by comparing the performance of inbreds used in making single crosses, with the results obtained when these inbreds are crossed with a common pollen parent, usually itself an inbred variety-cross.

Such a cross is called a *top-cross*. It is important that the top-cross parent has a broad genetic basis. After the more promising lines have been selected by the top-cross method, i.e. for their general combining ability, it is necessary to be able to identify the particular double-cross that will give the highest *specific combining ability*.

The most accurate method for achieving this is to ascertain the mean yield of the four single parent crosses, $A \times C$, $A \times D$, $B \times C$, and $B \times D$. ALLARD (1964) gives an example of the results when 10 single-crosses formed from 5 inbred lines were tested in this way (Table 3).

Table 3   Yields in bushels per acre of the 10 possible single-crosses among 5 inbred lines of maize. (From ALLARD, 1964; courtesy of Wiley & Sons)

Cross	Yield	Cross	Yield
(23 × 24)	41·7	(24 × 27)	72·1
(23 × 26)	62·6	(24 × 28)	69·3
(23 × 27)	70·8	(26 × 27)	64·2
(23 × 28)	64·4	(26 × 28)	60·4
(24 × 26)	65·6	(27 × 28)	59·6

Difference required for significance, 6·84

These data were then used to predict the yields of the 15 double-crosses that can be made up from the 10 single-crosses. For instance, the predicted yield of the double-cross (23 × 24) (26 × 27) was calculated as the average of the four single crosses (23 × 26), (23 × 27) (24 × 26) and (24 × 27), which is 67·8 bushels per acre. In a replicated trial the actual yield from (23 × 24) (26 × 27) was 68·8 bushels per acre. Yields from other double-crosses ranged from 56·0 to 71·1 bushels per acre and the predicted yields were remarkedly close to the actual ones in each case.

It has been established that the use of many inbred lines of diverse parentage is essential to the successful production of double-cross hybrids. Failure to use enough lines results in impoverishment of the genic content of the inbreds compared with the original open pollinated stock, i.e. heterozygosity must be maintained.

Production of seed in single-crosses is usually made by interplating two rows of the seed parent to one of the inbred pollinator. Similarly, with double-crosses six rows of the seed parent may alternate with two rows of the pollen parent.

The breeding of hybrid varieties of maize has been described in some detail to emphasize the unsatisfactory results from mere trial and error production of hybrids and the necessity of being able to predict the performance of outstanding hybrids to save costs.

Heterosis has been found in the crop plants, carrot, egg plant, marrow, onion, sorghum, tomato, and among flowering ornamentals, *Ageratum*, *Antirrhinum*, marigolds, *Petunia*, *Zinnia*. In England, the F_1 hybrid tomato Eurocross is widely grown, also Ware Cross. In the Brussels sprout the Japanese F_1, Jade Cross, has been a considerable success, as have also Japanese F_1 cabbages. In 1966 the F_1 sprout Avoncross was released by the National Vegetable Research Station. American F_1 onions have failed to maintain their early promise in England and seem largely to have been dropped.

Whether or not heterosis can be exploited commercially depends greatly on the ease or difficulty of emasculation and pollination. Both of these are problems with small-flowered species. In others, the issue may turn on the value of the crop, hence the price the grower is willing to pay for the expense of producing the seed. All glasshouse tomatoes in Europe must be produced by hand emasculation and pollination, but this is economic because of the very high gross return on the crop, £10,000 to £20,000 per acre. That the use of hybrid tomato seed is profitable is shown by the fact that more than 40 per cent of all glasshouse tomatoes grown in the UK are hybrids, and over 70 per cent in Guernsey.

An outstanding example of the use of male-steriles to produce hybrids on a large scale is that of pearl millet in India. This crop occupies some 29 million acres and has been bred by open pollination for more than 40 years. A crash programme was started in 1962 and by 1964 a hybrid had been produced which in trials at locations between 31°N. and 11°N. out-yielded local varieties by an average of 88 per cent. In 1965 this hybrid was recommended for release for the entire country, a notable success story by any standards (FREY, 1966).

4.5   Mutation breeding

As discussed earlier, natural gene mutation followed by gene recombination have provided the basic genetic variability upon which selection works with respect to the internal and external environments.

STADLER (1942) has shown that for different loci in maize the spontaneous mutation rate may vary from as much as 1 per 2,000 gametes to as little as 1 in 1,000,000, or less. From other workers' observations, and discounting extreme frequencies, we may assume that spontaneous mutation frequencies are commonly of the order 1 in 10,000 to 1 in 100,000 gametes, i.e. mutation is a recurrent phenomenon and, in a given species, mutants observed today have probably occurred many times in the evolution of that species.

We have also seen that the great majority of mutations are deleterious, often lethal. In other words, nature is, in a sense, profligate with mutations as well as plant progeny (cf. Darwin), hence natural selection of both is powerfully discriminate and the requirements for survival are extremely rigorous. Natural selection is, in fact, a very effective sieve.

When MULLER (1928) showed that some mutations induced in *Drosophila* by x-rays were indistinguishable from naturally occurring ones, plant breeders tended to assume that irradiation would provide a powerful tool for the induction of novel mutations of value to plant improvement.

Disillusionment followed fairly quickly: most induced mutations, like natural ones, proved deleterious and it was found that many of the mutant genes which might have improved commercially desirable characters already occurred in plant collections where these were adequate in variety. Moreover, we now know that such spontaneous mutations must have been, or became, moderately well adapted to their genomes to have survived at all, whereas new induced mutations would not 'qualify' under either natural or artificial selection, let alone satisfy the rigorous demands of commercial varieties.

We have approached the subject of induced mutation in this way because it provides a good example of how lack of a clear grasp of genetic principles (not understood in earlier decades) can result in much waste of breeding time and effort.

The most intensive work on mutation breeding is that of the Swedish breeders, which began shortly after Muller's discovery of the mutagenic action of x-rays, and has been continued until the present day.

To date, four varieties have been approved for release which originated as x-ray mutants: the white mustard Primex, (1950) the pea (*Pisum*), Strål (1954), and Pallas (1958) and Mari (1960) barleys. The barleys have been used as parents for further breeding, and two derivatives of these have been released. In Sweden, Norway and Finland, Mari barley is grown to a large extent as an early variety, ten days earlier than the original variety from which it was induced. Pallas has been widely grown in Sweden and Denmark, and fairly widely in England, but widespread infection by mildew in England in 1962–3 led to its being largely dropped by farmers in this country. Mari has not performed sufficiently well in England to warrant official recommendation. Milns Golden Promise barley, a gamma-ray mutant of Maythorpe, was the first variety to be accepted for plant breeders' rights. In has a very short stiff straw and has been grown to a small extent in England but proved susceptible to mildew.

In the USA, an x-ray mutant variety of peanut, NC4x, was released in 1959 after ten years of development. It outyields its mother strain (NC2, grown on 75 per cent of the peanut acreage of North Carolina), is comparable with the best varieties grown, has a thicker hull and is not so subject to pod cracks and the consequent kernel damage.

In Japan, a gamma-ray mutant variety of rice, Reimei, was released in

1966. The mutant was isolated from irradiation of the mother variety, Fugi Minori, in 1959, and shows high resistance to lodging, less variation in yield due to year and location than the parent, and has a better germination capacity and seedling growth under cultivation at low temperatures.

A special case is the use of mutation breeding to produce forms that do not exist in nature. All diploid sweet cherries are self-incompatible, a considerable disadvantage to the fruit grower. LEWIS and CROWE (1954) irradiated pollen-mother cells of superior varieties and used the mature pollen from the mother trees on others of the same incompatibility group. Some of the seedlings from these pollinations were self-fruitful and carried a mutant gene S_f (e.g. in S_3 S_f plants) and from these, homozygous forms (e.g. S_f S_f) are being bred. In this technique all non-mutant self-incompatible pollen grains are screened out by the style and only the mutant pollen (produced at low frequency in x-irradiated pollen) is able to effect fertilization.

A variety of agents has already been used in mutation breeding: alpha particles, neutrons, x-rays, gamma-rays, ultra-violet light, mustard gas, and various nucleosides, and it seems that the proportion of chromosomal disturbances and point (gene) mutations are positively and negatively correlated respectively with the energy dissipated by these mutagenic agents.

Mutagens which selectively produce point mutations would seem to be the ideal tools for unlocking the untouched store of potential variability in plant species, but when, and if, this situation is realized, one other factor of outstanding importance will remain to be considered. The natural store of variability which can be assembled in plant collections is potentially immense and in the end the choice between using natural or artificial mutations may well turn on the economic aspects of research and breeding, species by species.

To sum up. There has been a conflict of opinion about the value of mutation breeding, with those for and against its potential use arguing from theory and practice with some vigour, the theory not infrequently being unsupported by experiment and the practice not infrequently being based on inadequate information. Out of this confusion one conclusion clearly emerges. For the successful exploitation of induced mutation, two requirements must be satisfied: (a) detailed information on the identification, comparison and statistics of the quantitative and qualitative effects of mutagenic agents in different species and (b) rigorous selection of promising mutant characters for specific purposes and environments, just as with non-mutant breeding. A great deal of basic and applied research is now proceeding to evaluate the methods and consequences of mutation breeding. That it will take an accepted place among breeding techniques there is now little doubt. What is not yet clear is its relative importance among other techniques.

4.6 Induced polyploidy

Polyploids occupy so important a place among domesticated plants, it might be thought that the ability to create polyploids from diploids would provide such valuable advantages as to revolutionize plant improvement. This view, however, is immediately discounted when we consider the essential difference between established polyploids and newly created ones. The former have been exposed to the long-continued rigorous forces of natural selection and have been tried and tested for genetic and chromosomal balance and stability, fertility, vigour, etc. In contrast, the rawly new polyploid is untested by selection for any of these attributes and is characteristically less competent to survive, even when cossetted by man. To put it another way, the old genes and chromosomes are not adapted to the new, and therefore alien, internal polyploid environment which they themselves comprise. Nevertheless, as better understanding of genetic principles and mechanisms has grown, induced polyploidy can now be seen to offer promising means for the manipulation of genes and chromosomes towards the synthesis of novel and improved varieties.

Much work has been done in Sweden on autotetraploid varieties in the forage crops alsike and red clover. Fertility, though good in some strains, is less in the tetraploids than the diploids, though hay yields have been higher, chiefly because the plants recovered more quickly after grazing. Adverse characteristics include long styles, the inability of bees to reach the nectar in the flowers, and contamination of tetraploid stands by diploids. The best diploids have not always given the best tetraploids, i.e. the behaviour of the induced tetraploids is unpredictable. No tetraploid varieties have yet been released.

Summing up the prospects of utilizing induced polyploids, it is clear that new autotetraploids produced from a single diploid form should not be expected to be immediately successful. The way to success lies in producing a number of autotetraploids from genetically different diploid sources, followed by intensive hybridization and selection in large populations.

Some success has been achieved using triploids in commercial sugar-beet production (ELLERTON, 1967). The problem, however, of successful exploitation of triploidy has difficulties, the salient features of which are as follows. In general, triploids have larger roots than diploids, tetraploids smaller. Triploids yield the most sugar per acre. In Europe, the 'polyploid' (anisoploid) varieties grown are diploid-triploid-tetraploid mixtures, often in the approximate ratio of 40:60:10, produced by growing together diploid and tetraploid seed plants. Such varieties are widely grown in Europe and are predominant in EEC countries.

The only 'triploid' in cultivation is Triplex (syn. Trirave) and this occupies some 20 per cent of the British beet acreage. This is produced by growing together a diploid that is cytoplasmically male-sterile and a tetraploid pollinator. The seed is obtained from the diploid plants and is therefore of normal size and adapted to standard size ranges of precision drills.

Ideally, a high degree of cytoplasmic male sterility is necessary, not influenced by environmental conditions as is the case with much male-sterile material. The attainment of complete homozygosity in the parent used to maintain male-sterile lines is made difficult on account of beet being nearly self-incompatible.

In breeding for better male-sterile lines a further complication arose in that the tetraploid plants employed as pollinators had thicker and stronger anther walls which needed a drier atmosphere than the diploids before they burst and shed their pollen, thus reducing the efficiency of pollination. As a result, the percentage of triploids produced from the interplanting of diploids and tetraploids was found to vary from 90 per cent in favourable years down to 60 in unfavourable ones. The diploid component in the 'triploid' varieties came occasionally from male-sterile plants and from normal diploid beet, bolters, wild beet, red beet, etc. Thus, although the beneficial effect of triploidy *per se* was satisfactory, yields varied year to year, from very good to very bad, especially because of bolting.

Many individual lots of triploid test-hybrids have given phenomenal results, but reselection of tetraploids for easier pollen shedding, better performing male-steriles, and improved techniques for preventing contamination by pollen from outside sources, are all necessary before triploid sugar-beet can become as successful as its potential promises.

Sugar-beet affords a good example of the intricate problems which have to be solved by the breeder in the commercial exploitation of polyploidy.

4.7 Chromosome manipulation

Though aneuploids in themselves are of little interest for breeding purposes, they provide important means for the location of genes on chromosomes, the determination of linkage groups and the substitution of chromosomes of one species or variety for those of another (RILEY and LEWIS, 1966).

The manipulation techniques are based on the fact that non-disjunction at meiosis occasionally gives rise, in the simplest case, to one extra or one less chromosome in the complement, and these gametes, if viable, when united with a normal gamete in fertilization give diploid progeny of the constitution $2n+1$ (trisomic) or $2n-1$ (monosomic). Aberrant gametes or zygotes rarely survive in basic diploid types but are more likely to be viable in plants with polyploid ancestry, where chromosome replication covers the loss or gain of a chromosome.

Usually, the loss of a chromosome results in the aberrant form being morphologically distinct from the normal, and since any one of a set of chromosomes may be lost as described above, the different monosomics are also phenotypically different from one another. Similarly with tri-somics.

In the diploid tomato ($2n = 24$) all twelve trisomics have been made, with $24 + 1$ chromosomes. In the hexaploid wheat, Chinese Spring ($2n = 42$), the complete set of nullisomics has been obtained and the aneuploid forms of Chinese Spring have been used to produce complete aneuploid series in other wheat varieties.

Using techniques similar to those described above, different genes conferring resistance to stem rust have been located in nine varieties of wheat, and on the basis of this information a variety has been synthesized in which six genes found separately among the nine varieties have been combined in one variety, Thatcher.

In chromosome substitution, one or more pairs or chromosomes of one genotype are replaced by an equal number of pairs from another. This technique has been used to transfer a dominant allele for hypersensitivity to tobacco mosaic virus from *Nicotiana glutinosum* to tobacco, substituting one pair of tobacco chromosomes by a pair from *N. glutinosum* carrying the resistant gene. Similarly, rye chromosomes carrying desirable genes have been substituted for wheat chromosomes, and programmes for inter-varietal substitution have been drawn up.

4.8   Disease resistance

Perhaps the most intractable problem in plant improvement has been how to breed for resistance to the many pathogenic fungi that have, and can, ruin crops, sometimes catastrophically.

In many varieties resistance is known to be conditioned by a single gene, but duplicate, complementary and multiple alleles for resistance have also been found. Early hopes that factorial (Mendelian) genetics would permit rapid analysis and synthesis of resistant varieties were soon dashed; varieties resistant in one season were susceptible in another, or varieties resistant in one locality succumbed in others. Not until many intensive investigations had been made on the reproductive systems in fungi did the nature of the plant breeder's problem become evident, and adequate genetic studies become possible.

The pathogenic (and other) fungi exhibit a remarkable diversity of reproductive systems, some asexual, some sexual. The various mating systems, homothallic, heterothallic, the monocaryon, dicaryon and hetero-caryon incompatibility mechanisms, parasexuality, can, in the broad, be regarded as systems promoting outbreeding, and therefore conserving variability. It is now known that many species of pathogens comprise a great number of races (genotypes), e.g. several hundred in black stem rust of wheat, which differ in their physiological properties.

Physiological races are highly stable but, in the sexually producing fungi, genetic recombination may give rise to numerous other races as a result of the segregation of genes for virulence and non-virulence. A similar plasticity is found in many of the asexual pathogens. Thus, heterocaryosis has the ability to produce adaptive changes in response to alterations in the environment and to give rise to new pathogenic races (NELSON, WILCOXSON and CHRISTENSEN, 1955).

It will be realized from the outline given above that host-pathogen relationships are highly specific and highly numerous. The work of FLOR (1956) on flax has admirably demonstrated the gene-for-gene basis of the host-pathogen interaction in this species. Twenty-six alleles for resistance in flax have been located among five different loci; 2 at one locus, 3, 4, 6, and 11 at the other four loci, i.e. rust resistance is governed by multiple alleles and is inherited as a dominant characteristic. With one exception, virulence in flax rust (*Melampsora linii*) is inherited as a recessive character.

Singling out the host-pathogen relationships of flax rust presents, from a practical point of view, a deceptively simple situation when it comes to the general practice of breeding for resistance in the many crops subject to attack by pathogens. Obligate parasites, such as flax rust, cannot live apart from their hosts and as both must have evolved together, there is good reason for supposing that gene-for-gene relationships occur in other species. But genetical analysis in any species may be complicated by such factors as complementary genes, modifying genes, linkage, and polyploidy with respect to genes for resistance. Further, the capacity for resistance and the degree of pathogenic virulence in varying environmental conditions cannot be determined in the absence, or scarcity, of the pathogen and its specific physiologic races.

Because conditions outdoors are largely out of control, modern programmes of breeding for disease resistance increasingly necessitate the working out of host-pathogen relationships under controlled conditions in laboratories and special glasshouses, where plant genotypes can be positively identified, also the different physiologic races, and both maintained for infection tests. This, in turn, demands close co-operation between plant breeders and pathologists.

Returning to the question of breeding methods and results, simple selection of plants resistant in the field has been successful in cabbage, sorghum, sugar-beet and alfalfa. In some cases, genes for resistance have not been found among cultivated varieties, but in related species. In such cases, the backcross method (p. 23) has been used to transfer the gene for resistance into established varieties using these as recurrent parents.

A current example of the transfer, by backcrossing, of resistant genes in species to cultivated varieties, is that of tomato, *Lycopersicon esculentum*. Resistance to tomato mosaic virus, which causes considerable loss in yield (10–20 per cent) and quality of fruit in the glasshouse tomato crop, has been found in *L. hirsutum*, *L. peruvianum* and *L. chilense*. Resistance to the

Plate 5 Quantitative inheritance of doubleness of flowers in the garden dahlia showing the parents (perfect single and perfect double) and range of variation in a sample of F_1. Note parental types are not recovered. (CRANE, M. B. and LAWRENCE, W. J. C., 1952. *Genetics of Garden Plants*, Macmillan, London.)

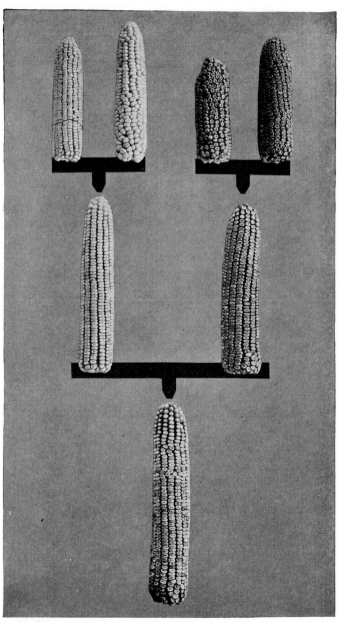

Plate 6 The result of bringing together four inbred lines of maize, by three crosses. Note hybrid vigour. (With acknowledgements to Dr. D. F. Jones.)

complex disease, brown root rot, has been found in *L. hirsutum*, and resistance to *Fusarium* wilt in *L. pimpinellifolium*. Some of the genes conferring resistance have been identified and when the genetical situation has been established and the physiologic races identified, it is hoped to combine in the best commercial varieties the several genes conferring resistance to the races of any one of these diseases, with similar genes conferring resistance to the other diseases.

The cost of steam sterilization of glasshouse soil, of chemical sprays and the various cultural devices aimed at minimizing disease, plus the loss of money from diminished fruit yields and quality, are a recurring addition to the cost of tomato production. To lessen and eventually eliminate this burden can only be achieved by biological control, i.e. by the plant breeder producing varieties that make unnecessary the expensive and second-best methods of disease 'control' currently practised.

In wheat, resistance to bunt fungi (*Tilletia caries* and *T. foetida*) has been studied in 11 different varieties, using 26 physiologic races (BRIGGS and HOLTON, 1950). It was found that each gene controlled several races of both *T. caries* and *T. foetida*, such that the 25 races could be divided into 8 groups for pathogenicity. The outcome of this investigation was to establish that resistance to all known races of bunt in wheat could be controlled by the joint action of three, identified, genes.

In England, one or two multi-resistant cereal varieties have been grown in recent years. Cappelle-Desprez, the widest grown winter wheat variety, carries two genes for resistance to three races of yellow rust (*Puccinia striiformis*), and has partial resistance to a fourth race. Impala spring barley has proved very successful and has mildew resistance derived from two sources, the German variety Wisa and *Hordeum spontaneum*. Rothwell Perdix winter wheat was claimed to be a multi-line variety, but proved to be susceptible to a new race of *P. striiformis* in 1966. There did not appear to be any differential susceptibility between the lines.

The use of resistant multi-lines is likely to increase in the attempt to reduce the prevalence of fungal diseases, but to accomplish this successfully several requirements must be met: (a) a number of different sources of resistance must be repeatedly backcrossed to one of the best commercial varieties, (b) varietal lines must be produced which are highly uniform with respect to cultural, developmental and quality characteristics, and (c) to meet a range of agricultural conditions the programme must be extended to include several commercial varieties, within each of which resistant lines have been bred as above.

Although fewer genetic studies have been made on bacteria, nematodes and insect pests, there is evidence to suggest that they, like the fungi, can produce numerous physiologic races.

It will be evident even from the few aspects discussed above that breeding for disease resistance calls for rigorous techniques, based firmly on sound genetical principles (see p. 37; also ALLARD, 1964).

4—P.B.

4.9 Environmental control

A handicap in much plant breeding is the interval between generations, often a year. This can be reduced in some instances when seed can be harvested in six months or so from sowing, by growing one crop through the summer, say in a temperate zone, and then raising a second generation in another part of the world where summer conditions prevail. In this procedure, the breeder must be on his guard against such intrusive factors as the effect of photoperiodism, and the complications that may arise in making selections in one clime for eventual use in another.

In the winter season, time can also be saved by using heated glasshouses, in which temperature is controlled and growth thereby secured when it would be impossible outdoors. The heated glasshouse has other advantages for winter work. The duration and intensity of natural light can be supplemented by artificial light, to accelerate growth photosynthetically and to induce flowering photoperiodically. To these benefits can be added those obtained from the use of CO_2 (now a common practice with commercial glasshouse crops) to increase growth rate still further and shorten the time to flowering. The ultimate refinement in this direction is the use of cabinets or chambers in which light duration and intensity, temperature, humidity, and CO_2 concentration, are controlled with high accuracy. Assemblies of such chambers together with full ancillary facilities are known as *phytotrons*.

Controlled environments can be used for shortening the time between generations and for making preliminary assessments of the response of specific genotypes to specific environments. The employment of controlled environments is becoming widespread. It may be confidently expected that before long chemical (e.g. hormonal) control of growth and flowering will further supplement the benefits afforded by phytotrons.

4.10 Distribution and maintenance of new varieties

The benefits promised by a new variety cannot be realized until (a) enough seed has been produced to permit commercial growing over the whole area to which the variety is adapted, and (b) provision has been made to maintain the purity of the variety.

The details for distribution may differ somewhat between different countries, but in essence the stages are common to all. The primary function of the breeder is to develop new varieties and carry out initial small-scale seed increase. Such seed is called *breeders' seed*. The second stage is the multiplication of breeders' seed to give *basic seed* followed by multiplication of this to provide *commercial seed* for farmers and growers.

In the UK the National Institute of Agricultural Botany is responsible for the production of basic seed of successful varieties bred at the government breeding stations. The multiplication of seed from private or foreign breeders is controlled by the commercial firms involved directly or as agents of foreign breeders.

In Britain, new varieties can be patented by breeders under conditions, specified by the Plant Varieties and Seeds Act, 1964. At present, the scheme applies to wheat, barley, oats, potatoes, rhubarb, strawberries, apples, pears, dahlias, roses and delphiniums, and the plant breeder's rights run from 15 to 25 years, according to the species. During the period the rights remain in force the breeder has the exclusive right to reproduce and sell material in the UK. Similar schemes operate in other countries.

Breeding Principles 5

5.1 Fitness

In the preceding chapters we have discussed the bricks and mortar of plant breeding; the raw material for the building of new varieties. We have now to consider the architecture of breeding, the factors inherent to efficient and rewarding design, a subject given all too scant attention in the past.

Darwin's phrase 'survival of the fittest' has become so much a part of the English language as to seem self-evident in reference to any struggle or competition for existence. The phrase is, in fact, so general as it stands as to have little meaning. Fit to survive? Variety or species? Individual or population? In a static or changing environment? For how long: a month, year, century or millennia? Under what influences: light, temperature, water supply, nutrition, space, man? And what are the components of the organism which bring about its fitness or survival, and what are their interactions?

We recognize almost intuitively that competition for the component factors of the environment external to the plant must have vital consequences for its survival. What has to be understood, however, is that ultimate survival starts with the components of the *internal* environment, with the nature and sum of its hereditary material, with the genotype.

In discussing fitness in wild populations we must distinguish two different categories: fitness of the individual, fitness of the species. An individual highly adapted to flourish in one specific environment has no future if that environment changes to a significant degree. And a species composed of such individuals also has no future. Over-specialization, regardless of other factors, tends to extinction in the evolutionary race. This we all know.

What are the genetic devices that confer rigidity or adaptability on an organism or species? In a hypothetical wholly inbreeding population consisting of many homozygous lines, selection will favour the more balanced homozygotes at the expense of the less balanced ones. The favoured individuals will reproduce themselves and, so long as the environment is stable, the population will consist of the *optimum* phenotype (which may be the expression of more than one genotype). Variability is restricted, indeed frozen.

If the environment changes a little, then all that happens is that other homozygous lines are selected which are better adapted to the new environment, and a new optimum phenotype may be established as before. Variability, however, cannot exceed that carried *between* lines and is unable to meet the requirements of a major change in the environment. Such a population shows high immediate fitness but little flexibility and the inbreeding system becomes a handicap when the environment changes. Occasionally, mutation will introduce a new gene, but this will rapidly be fixed in one line and the *status quo* is the same as before. By and large, such an inbreeding system would be in an evolutionary dead end. There is no variability upon which selection can work and the potential for adaptation is at a minimum. The one escape from this genetical impasse is for some outbreeding to occur and it is becoming apparent, in fact, that many so-called inbreeders outcross to a surprising extent, and then recombination and segregation, the most powerful mechanisms for the preservation of variability and adaptability, can come into play for a while. Incidentally, outbreeding can be induced artifically in inbreeders, when male-sterile forms are available from which to save seed, the product of random mating.

In an outbreeding system, the heterozygotes will always segregate a proportion of individuals in which some alleles are homozygous (i.e. extremes can never be lost), therefore the population must always vary around the optimum phenotype. Selection favours balanced genotypes of *intermediate* phenotypic expression. The immediate fitness of the outbreeding species is lower than in the homozygous inbreeding species, but adaptability is greater because of the flow of variability inherent to the outbreeding system. Thus, the genetical constitution of the outbreeding population can, and will, change to meet the demands of new conditions. The future is not an evolutionary cul-de-sac. Adaptation is optimal and variability free.

Adaptability must not be excessive, of course, and here other regulating and balancing systems come into operation, such as gene linkage. Selection works against the less well balanced combinations of genes and in consequence those conferring balance will tend to become linked in inheritance with the result that the spread around the optimum phenotype will be less, but short-term fitness higher. Not infrequently genes of vital importance to some aspect of plant morphology or physiology will become so closely linked under selection as to comprise a *supergene* in which crossing-over and recombinations are very rare events.

A clear example of this is found in the diploid outbreeding genus, *Streptocarpus* (Cape Primrose). Eighty years ago, three species were received at the Royal Botanic Gardens, Kew, and intercrossed to combine the flower colours from which were later bred the hybrid cultivars now comprising the ornamental glasshouse forms. The three species exhibited different flower colour patterns, and genetic analyses (LAWRENCE, 1957) revealed that the four genes determining colour pattern were, as would be expected, correspondingly different between the species, viz *S. rexii FFüBBHH*, *S. parviflorus ffiibbHH*, and *S. dunnii FFIIbbhh*.

Genetic analysis of the many different garden forms, derived by the crossing of these species and subsequent selection by breeders, showed that the only viable gametes were those of the same constitution as in the species. Thus instead of the 81 genotypic classes expected from the independent inheritance of 4 pairs of alleles, only 6 classes are found. Each of the parental combinations of 4 genes, (*FiBH*, *fibH*, *FIbh*) is inherited as 1 unit, a supergene. Practically all other combinations are lethal, indeed only 8 anomalous genotypes were found in many thousands of cultivated plants, and 3 of these originated by mutation at the *I* locus. The 3 supergenes have remained virtually intact through 80 or more generations of intensive breeding.

What is of additional interest is that the many species of *Streptocarpus* apart from the three mentioned, show other combinations of the linked alleles. It seems, therefore, that these flower patterns have a high adaptive value in preventing the break-up of processes and characters vital to the survival of *Streptocarpus* species in the wild, the different patterns being adapted possibly to meet the requirements of different insect pollinators in different climatic and topographical regions. Variability with respect to flower pattern genes is tied up for the survival of the species.

The restriction of recombination in outbreeders can be achieved by several means: by genetic control of the number of chiasmata; by localization of chiasmata to a specific region or regions of a chromosome; by inversion of short chromosome segments within which as a consequence recombination is suppressed (this is a mechanism giving rise to supergenes). One outcome of this interplay of the opposing mechanisms is that while segregation and recombination produce superior combinations of genes, these combinations are so desirable as to make further recombinations undesirable for the time being. Balanced integration is itself under genetic control.

These genetic components of fitness have been presented to show, albeit incompletely, the background which the breeder must not only be aware of, but understand and never overlook. Variability is, so to speak, the life blood of a species; controlled so as not to be released disruptively, stored to anticipate the dictates of environmental change. The store is not one in which every gene is unique. For example, in the course of evolution, many genes are duplicated with respect to their functions (especially polygenes).

The greatest store of variability is found, not among major genes, but polygenes. Polygenic inheritance allows the storage of enormous reserves of variability, in that adaptive response—the reaction between variability and selection—permits the fitting together of favourable complexes from genes which individually are not particularly favourable. New variations which may have a deleterious effect in the combination and environment in which they occur, when carried through a population in recessive condition may appear in a more suitable environment and a more suitable combination where they can increase.

The store of variability is greatest when numerous polygenic alleles are heterozygous, as will readily be seen from the following example. Consider two homozygotes, *AA* and *aa*. When these are crossed, the hybrid *Aa* upon selfing (or crossing with its like) will give, by segregation, *AA*, *Aa* and *aa* forms in the next generation, i.e. the heterozygote has latent or *potential* variability which is made *free* in its offspring. And what is true of one pair of alleles is also true for many, for polygenes. In outbreeding populations part of the variability is potential (in heterozygotes) and part free (in homozygotes). In closely inbreeding populations all the variability has been freed and none is potential since there are no heterozygotes.

It may be noted here, that identical phenotypes may be produced by a large variety of genotypes, in response to intangible differences in the environment, i.e. the genetic basis of form may change though form itself does not. Thus, even different farming systems can impose their own selective effects via the polygenes on the primary climatic adaptation of crops (COOPER, 1965), and permit the emergence of agro-ecotypes.

So far we have been discussing fitness, largely in general terms, as between inbreeders and outbreeders, genotype and external environment. The notion of fitness, however, is just as pertinent to the hereditary mechanism, for the chromosomes and genes are themselves under the control of the genotype: mutation rate, frequency of recombination, inversions, chromosome size and number, and so forth.

Each gene has multiple effects on the phenotype, physiologically and morphologically. Major genes are merely those having, among others, outstanding effect which captures attention. Every new genotypic variation of importance throws the organism into a new environment. Every gene change demands others to act in concert, i.e. each component of the genotype must be adapted to the others, to the total internal environment.

In toto, therefore, the genotype comprises a dynamic, co-adapted polygenic system, sensitively balanced to the requirements of the genotype as such, to phenotypic expression, to adaptation of both the individual and the species: and the genes are selected as an integrated whole. 'The genes are like members of a legislature in being subject to the laws they enact as a body' (DARLINGTON, 1939).

When we turn to domestic crops, we find the criteria of fitness are different from those for wild plants. The fittest individual, crop, or species, is that in which all the desirable attributes for commercial growing are brought together in adequate degree, ideally to the maximum possible. Nature's aim is survival in space and time. The grower's aim is solely productivity. For instance, to meet man's requirements maize has lost the capacity of its progenitors for seed dispersal; in some forage grasses reproduction has been deliberately delayed and thereby reduced; many sterile horticultural plants must be propagated asexually. 'Where nature induces variety and relational balance within adaptive communities, the farmer favours simplicity and uniformity, stability and repeatability.

Adaptability under domestication has, then, a specific and restrictive meaning. It is the response of an organism to a range of environments in terms of economic output' (FRANKEL, 1966). In other words, fitness in domestic crops is artificially restricted to unnatural requirements, with only a limited ability to adapt in the (not very) long term changes in the total environment, natural or man-made. Gene erosion and genotypic rigidity tend to be the order of the day.

5.2   Adaptability

In Chapter 2, the advantages of uniformity in commercial crops was briefly mentioned: uniformity in relation to the requirements of husbandry, marketing, canning, retail sales, etc. This call for high uniformity is relatively new and is spreading under the demands of mass production, processing and consumption. To this situation must be added another; the move, under economic pressures, to larger units of production, whether farms, market gardens or glasshouse crops. This is not infrequently accompanied by a reduction in the number of varieties grown in a given crop: man selects the most favourable varieties for his purposes to the exclusion of the less favourable ones. As we have seen, a danger in this situation is that the gene pool of a cultivated species may be depleted and, indeed, genes lost if precautions are not taken.

This leads to the question of adaptability of advanced economic crops. Clearly, when greatly increased food production is urgently necessary in the vast areas and for the vast populations of undeveloped countries, it should be quicker to start with crop varieties already well proved and successful in developed countries, than to start programmes with the limited, and perhaps inferior, plant resources of the undeveloped areas, always supposing the advanced crop varieties can be adapted to the environments under consideration.

In view of what has been said about the conservation of uniformity in crops under high agricultural standards, we must ask the question, what degree of adaptability remains in these crops ? Does it exist, and how can it be measured ?

This is a new field of investigation requiring techniques just beginning to be originated and evaluated. Only one example will be given here of the approach to obtaining reliable quantitative measurements of adaptability. FINLAY and WILKINSON (1963) in Australia used a representative range of several hundred barleys from most parts of the world and tested them in a variety of environments. Two indices were derived which in combination provided a measure of the adaptability of individual varieties over a range of environments, and of the degree of adaptation to particular environments. The first index is a measure of stability and is the regression coefficient of a variety on the mean of all varieties. The second index is the mean performance of the variety over all tests (Fig. 5–1).

This method provides a simple scaling of adaptability over specified sites and seasons, provided the test collection and the range of environments are broadly representative.

Important as is the rapid exploitation of adaptability inherent in economic crops, it but touches the fringe of a very much wider issue, one upon which the whole future of plant breeding depends, namely, the relation of adaptation to adaptability. The simplest way of expressing this relation is to say that adaptation and adaptability are in conflict. Closeness of adapta-

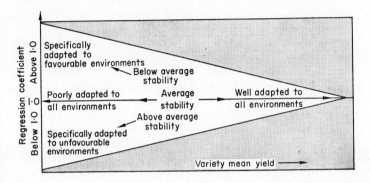

Fig. 5–1 How Finlay and Wilkinson propose that the adaptability of a crop plant variety should be assessed. The performance (yield per acre) of a range of varieties in a range of environments is measured. The mean yield of the variety in question locates it along the horizontal axis of this diagram. Its position along the vertical axis is computed by comparing the variety's yield with those of other varieties, to arrive at a 'regression coefficient'. (FRANKEL, O. H., 1966. *New Scientist*, **31**, 144, after FINLAY, K. W. and WILKINSON, G. N., 1963. *Aust. J. agric. Res.*, **14** (6), 742–54.)

tion implies restriction of adaptability. As we have seen, advanced agricultural crops are characterized by their adaptation to rigorous agricultural (and other) requirements, by the culling out of variability to the point where the genetic base (gene pool) of a crop species is so narrow that adequate adaptation to changes in the environment (e.g. new pathogens) is unlikely.

The reason for this state of affairs is due partly to the influences of historical events and partly to the neglect of genetical principles. The first impact of Mendelian genetics was, naturally, to focus attention on factorial breeding. Where a popular crop variety was deficient in a desirable character, then a major aim of the breeder was to breed this gene into the established variety, by backcrossing and, often, pure lining, the consequence of which in both cases was to reduce variability. Immediate fitness was sought to get immediate improvement in yield or some other obviously desirable character. Moreover, the wide range of variation in primitive crops has

long since become unacceptable to modern farming and growing; uni
formity in a few (relatively) superior varieties of a species became a high
priority, thus still further depleting the species of its genic content and
capacity for variability and adaptability. The European potato and glass
house tomato are classical instances of species which, when first introduced
from the wild, were but fragments of the wild populations, and ever since
breeders have been trying to ring the changes on cultivars having a narrow
genetic basis, too narrow to permit of any real improvement by breeding.

Nowhere has the heavy hand of factorial breeding led to such neglected
opportunities as in breeding for disease resistance. The facts outlined in
Chapter 1 have been available for years; nevertheless, many plant breeders
continued, and continue, practising 'a cat and mouse game with rapidly
evolving pathogen populations, seeking only a step by step evasion of each
new . . . race' (BENNETT, 1965). 'It is astonishing,' observed BECKER
(1959), 'that after 30 years of development in which genetics has come to
regard evolution as a problem of the genetics of populations, and has re
jected the misleading simplicity of the early explanations offered by fac
torial genetics, this same development has passed plant breeding by.'

The overwhelming importance of breeding for disease resistance is
amply demonstrated. In the 1947–8 wheat-stem rust epidemic in New
South Wales, the loss represented food for 3 million people for one year.
In Africa, annual losses of sorghum and millet have equalled not harvesting
these crops from an area of 5 million acres. Leaf spot of bananas reduced
the Mexican production of the Gros Michel variety by 50 per cent in two
years. In the USA alone, the cost of crop losses due to disease during the
period 1951–60 has been estimated to be more than 3,250 million dollars.

Factorial breeding for disease resistance is often at a discount because of
the narrow genetic basis of many crop plants. Lacking polygenic variation
such crops are pecularly vulnerable to pathogenic attack. Adaptability is
severely limited because adequate variability has been bred out, or was not
there to begin with.

Gene erosion by monogenic breeding and the replacement of local
varieties by highly uniform crops threatens 'all the plant breeding work
carried out by nature over thousands of years' (PAL, 1960). For instance
great acreages in Turkey have been planted to a single variety of flax and
'from one end of Cilicia to the other I could not find a single indigenous
variety although this very area had at one time been a centre of diversity for
flax' (HARLAN, 1956). This replacement of local varieties by highly uniform
varieties is steadily progressing in remote regions, e.g. Ethiopia.

5.3   Conservation of genetic variability

It is abundantly clear from the many examples that can be quoted that
the conservation of genetic variability is a prime and urgent necessity for
plant breeding in the future. Primitive cultivations are characteristically

variable: advanced cultivations characteristically uniform. And the rich variability of the primitive crops in undeveloped regions is fast being ousted by the 'improved' cultivars from the agriculturally advanced nations.

One solution proposed for the solving of this problem is the establishment of national and world collections of living crop varieties. Such collections have already been established in a number of countries: at the Institute of Plant Industry, Leningrad, over 160,000 living specimens are grown, excluding fruit trees; over 23,000 species are held at Gatersleben in Germany. MAO (1959) lists a total of nearly 400,000 strains, varieties and cultigens in world collections.

The value of collections, however, depends on their comprehensiveness and constant supplementation; mere numbers do not imply that the genetic base is as broad as is necessary. Further, breeders' collections are traditionally regarded as sources of 'characters', usually controlled by major genes, which may be extracted by backcross techniques, a kind of naïve lucky dip for the apparently obvious prizes.

The demand from the breeder 'is more intricate than the classical hypothesis would lead us to believe. It is not isolation of some one genotype which engenders superior qualities in its carriers. It is rather the construction of a new balanced gene pool of the population in such a way that the processes of Mendelian segregation and recombination usually yielded highly productive genotypes and phenotypes' (DOBZHANSKY, 1955).

An alternative proposal to the establishment of living collections alongside the breeders who use them is to set up the collections in the existing centres of diversity of the crops themselves: potatoes in the Andes, rice in India, flax in Turkey, and so on.

Neither solution, however, meets the basic requirements of mass reservoirs of variability. Collections of cultigens undoubtedly have their uses as breeding material but it seems they ought to be supplemented by varied hybrid populations maintained in bulk under minimal artificial selection (SIMMONDS, 1962). To this end, the FAO has recommended the establishment of International Crop Research Centres *within* the centres of natural genetic diversity, i.e. the setting up of areas of genetic conservation, where the evolutionary potential of local population-environment complexes can be preserved. One such centre has been established at Izmir in Turkey for the study of Mediterranean and Near East gene centres.

This is a far cry from the notion of plant improvement by simple manipulation of major Mendelian characters; as different in concept as Dalton's notion of the atom as a billiard ball compared to the fine dynamic structure of the atom revealed by modern physics. Knowing how to use the Mendelian bricks and mortar is essential to the fabrication of improved varieties, but the overriding issue is that of the design of the structure of plant breeding, the architecture of plant improvement. In practical terms, the aim should be to preserve genetic variability within the frame of phenotypes meeting modern agricultural standards.

An example of a current potato breeding programme will serve to emphasize the importance of this concept. European and North American varieties of the Tuberosum group of potato are tetraploids believed to have sprung from very limited introductions of the tetraploid Andigenum group, beginning in the sixteenth century and continuing intermittently until the present time. It is probable that the variability was a very small fraction of that which exists at the tetraploid level in South America.

The main object of European selection was tolerance of long days (Andigenas are often short-day plants which rarely tuber in the latitude of England). If variability in Tuberosum is limited, then crossing the best Tuberosum varieties with others from the Andigena group and selecting for long-day tubering should greatly increase variability, hence also the possibility of selecting for increased yields. In short, it is feasible that the Tuberosum group can be *reconstructed* using Andigenum stock.

This work has been in progress (SIMMONDS, 1966) for a bare five years but already yields have been increased by 50 per cent, the new varieties show marked heterosis, and some are highly resistant to potato blight (the resistance comes from Andigena).

That such conservation is an urgent necessity is well recognized by a number of plant breeders but by all too few politicians and governments, who in fact hold the future of world crop production in their hands. For instance, the entire acreage of the grass, *Digitaria decumbens*, in Central America and the West Indies originated from a single clone introduced in 1940; and only a handful of parent wheat lines have contributed to the gene content of the Australian wheats. Again, only a few hundred of inter-related genotypes comprise the cultivated bananas and sugar-canes.

The great majority of the trees of *Coffea arabica* in South America are derived from a few seedlings produced from a single mother tree in the Amsterdam Botanic Garden early in the eighteenth century. The coffee industry in Ceylon was built up in exactly the same way as that in South America, and in 1860 a leaf rust disease completely wiped it out. Should the same fungus disease get into South America, its coffee industry might well be destroyed.

5.4   Centres of diversity

Mention has been made of the increase in the collection of plants from Neolithic times to the beginning of this century. The salient feature of plant collection all through this time was its haphazard nature: travellers came across new or desirable species, varieties and cultivars and took them home. Haphazardness was also the mark of planned expeditions to the Himalayas, Far East, tropical rain forests, Australia, etc.; plant hunters

went to see what they could find to send or bring back, botanists to add to
the national herbaria.

One man came to see that a matter of great issue lay behind the collection of plants, namely Nikolai Vavilov, agricultural botanist, geneticist and
head of the Soviet Organization for Plant Breeding. Between 1923 and 1933
Vavilov sent out expeditions to 60 countries to collect cultivated plants
that might be useful in supplementing the collections accumulated since
1916 at Leningrad. His studies of numerous cultivars led him to recognize
ten main centres of diversity of the world's most important crop plants:
Abyssinia, Mediterranean, Near East, Central Asia, Indo-Burma, Indo-
Malaya, China, Central America, Eastern South America, and Western
South America. Each centre was characterized by a large number of cultivated forms of a given crop. Vavilov put forward the proposition that the
greater the number of forms of a crop within a centre, the greater was the
antiquity of that crop. The crop probably originated in that centre, hence
the centres of origin of species coincide with the areas where the greatest
diversity exists in the species.

On the basis of more extensive knowledge acquired in recent years, of
the distribution of variability in cultivated species, details of Vavilov's
original proposals have been criticized and modified; nevertheless, his
centres of diversity, perhaps better described as centres of development,
have turned out to be fertile collecting areas and still remain promising ones
for future collections. Vavilov's monumental contribution to an understanding of the origin and utilization of wild species and old and new
cultivars lay in the manner in which he directed attention to aspects of
plant collection and plant breeding not previously appreciated.

Following Vavilov's lead, a number of workers have located and investigated areas of diversity, and the conditions associated with these have been
examined in some detail. HARLAN (1959) has described gene micro-centres
in which are concentrated enormous variability'. These micro-centres are
associated with habitats recently disturbed by man and the presence of
wasteland species. They are also characterized by extensive *introgression*.
Introgression (ANDERSON, 1953) is the 'infiltration of the germ plasm of
one species into that of another', and results from hybridization and recurrent backcrossing (p. 24) to the parents.

An example in which introgression has played a part is described by
Harlan and refers to a typical wheat field in Thrace which 'contained a
wonderful mixture of forms. Around its borders, in the weed rows, the
roadsides, the waste spaces and to some extent in the fields themselves the
wild wheat relatives are found in abundance'. In such centres interbreed-
ing and evolution are proceeding at a rapid pace *now*. The importance
of the study of the origin and evolution of cultivated species, initiated by
Vavilov, can hardly be over-emphasized in relation to the conservation and
utilization of variability which must be the basis for modern plant
breeding.

5.5　The future

We have now, all too briefly, surveyed the whole field of plant breeding
the influence of primitive man on plant populations and some of the factor
in plant domestication; man's goals in manipulating plants for his ow
ends; the methods (tactics) breeders have used to achieve advances o
limited fronts; an analysis of the genetic structure of wild population
which has enabled them successfully to survive and evolve.

Our remaining task is twofold: to evaluate the past work of the breede
and to indicate what the future requires of him. These two aspects can b
succinctly summarized. In the past the breeder has been concerned wit
tactics. In the future he must concentrate on *strategy*. The strategic
requirements are abundantly clear: the breeder must adopt or adapt th
basic strategies and tactics nature has so successfully employed in the evolu
tion of plants, and which he ignores at his and the world's peril.

The field of operation is exceptionally broad and demands the balance
exploitation of all means and devices in accordance with the requirement
of the plant environment, internal and external, natural and artificia
short-term and long-term.

The urgent demands of a food-hungry world may, for some years t
come, demand crash programmes for quick results, but expediency shoul
not divert the breeder's attention from the ultimate goal in plant improve
ment: the selective conservation and utilization of genetic variability fc
the welfare of mankind.

References

ALLARD, R. W. (1964). *Principles of Plant Breeding*. Wiley, New York.

ANDERSON, E. (1949). *Introgressive Hybridization*. Wiley, New York.

ANDERSON, E. (1953). Introgressive hybridization. *Biol. Rev.* **28**, 280–307.

BECKER, G. (1959). Darwin und die Pflanzenzüchtung. *Ber. Vortr. dt. Akad. LandwWiss.* 4te Festsitzung, 99–113.

BENNETT, E. (1965). Plant introduction and conservation. *Scottish Pl. Breed. St. Rec.* 1965, 27–113.

BRIGGS, F. N. and HOLTON, C. S. (1950). Reaction of wheat varieties with known genes for resistance to races of bunt. *Agron. J.*, **42**, 483–486.

COOPER, J. P. (1965). The evolution of forage grasses and legumes. In *Essays on Crop Plant Evolution*, ed. J. B. Hutchinson, 142–165. University Press, Cambridge.

CRANE, M. B. and LAWRENCE, W. J. C. (1952). *Genetics of Garden Plants*. Macmillan, London.

DARLINGTON, C. D. (1939). *The Evolution of Genetic Systems*. University Press, Cambridge.

DARLINGTON, C. D. and MATHER, K. (1950). *The Elements of Genetics*. Allen & Unwin, London.

DARLINGTON, C. D. and WYLIE, A. P. (1956). *Chromosome Atlas*. Allen & Unwin, London.

DOBZHANSKY, TH. (1955). Concepts and problems of population genetics. *Cold Spring Harb. Symp. quant. Biol.*, **20**, 1–15.

ELLERTON, S. (1967). Private communication.

FINLAY, K. W. and WILKINSON, G. N. (1963). The analysis of adaptation in a plant breeding programme. *Aust. J. agric. Res.*, **14** (6), 742–754.

FLOR, H. H. (1956). The complementary genic systems in flax and flax rust. *Adv. Genet.*, **8**, 29–54.

FRANKEL, O. F. (1966). Adaptability of crops. *New Scient.*, **31**, 144–145.

FREY, K. J. (1966). *Plant Breeding*. Iowa State University Press.

HACKBARTH, J. and TROLL, H. J. (1956). Lupinen als körnerleguminosen und futterpflanzen. *Handbuch der Pflanzenzüchtung*. ed. H. Kappert and W. Rudolf. 2nd. ed. **4**, 1–51. Pavey, Berlin and Hamburg.

HARLAN, J. R. (1956). Distribution and utilization of natural variability in cultivated plants. *Brookhaven Symp. Biol.*, **9**, 191–208.

HARLAN, J. R. (1959). Plant exploration and the search for superior germ plasm for grasslands. *Publs Am. Ass. Advmt Sci.*, **53**, 3–11.

HELBAEK, H. (1959). Domestication of food plants in the Old World. *Science*, **130**, 365–372.

HUTCHINSON, J. (1965). *Essays on Crop Plant Evolution*. University Press, Cambridge.

KAKIZAKI, Y. and KASAI, T. (1933). Bud pollination in cabbage and radish. *J. Hered.* **24**, 359–360.

LAWRENCE, W. J. C. (1957). *Practical Plant Breeding*. Allen & Unwin, London.

LAWRENCE, W. J. C. (1957). Studies on *Streptocarpus*: IV Genetics of flower colour patterns. *Heredity* **11**, 337–357.

LAWRENCE, W. J. C. and SCOTT-MONCRIEFF, R. (1935). The genetics and chemistry of flower colour in *Dahlia. J. Genet.*, **30**, 155–226.

LEWIS, D. and CROWE, L. K. (1954). Self fertility in fruit trees. *Discovery*, **15**, 371.

MANGELSDORF, P. C. (1965). The evolution of maize. In *Essays on Crop Plant Evolution*, ed. by J. B. Hutchinson, 23–49. University Press, Cambridge.

MAO, Y. T. (1959). A preliminary list of crop plant collections and their custodians. *Pl. Introd. Newsl.*, **6**, 1–124.

MULLER, H. J. (1928). The measurement of gene mutation rate in *Drosophila* and its high variability and dependence upon temperature. *Genetics*, **13**, 279–359.

NELSON, R. R., WILCOXSON, R. D. and CHRISTENSEN, J. J. (1955). Heterocaryosis as a basis for variation in *Puccinia graminis* var. *tritici*. *Phytopathology*, **45**, 639–643.

PAL, B. P. *et al.* (1960). A study of survival in a mixture of fourteen varieties of wheat. *Indian J. Genet. Pl. Breed.*, **20**, 107–112.

RILEY, R. (1965). Cytogenetics and evolution of wheat. Essay in *Crop Plant Evolution*, ed. J. B. Hutchinson, 103–118. University Press, Cambridge.

RILEY, R. and LEWIS, K. R. (1966). *Chromosome Manipulations and Plant Genetics*. Oliver and Boyd, Edinburgh.

RYBIN, V. A. (1936). Spontane und experimentall erzeugte Bastarde zwischen Schwardorn und Kirschpflaume und das Abstammungsproblem der Kulturpflaume. *Planta*, **25**, 22–58.

SENGBUSCH, R. VON and ZIMMERMANN, K. (1947). 20 Jahre Süsslupinenforschung and Züchtung in Deutschland. *Forschn u. Fortschr.*, 21–23, 249–255.

SIMMONDS, N. W. (1962). Variability in crop plants, its use and conservation. *Biol. Rev.* **37**, 422–465.

SIMMONDS, N. W. (1966). Studies of the tetraploid potatoes, III. Progress in the experimental re-creation of the Tuberosum Group. *J. Linn. Soc.* (Bot.) **59**, 279–288.

SPRAGUE, G. F. (1966). Quantitative genetics in plant improvement. In *Plant Breeding*, ed. by K. J. Frey. Iowa State University Press.

STADLER, L. J. (1942). Some observations on gene variability and spontaneous mutation. *Spragg Memorial Lecture in Plant Breeding*. Michigan State College.

WILLIAMS, W. (1964). *Genetic Principles and Plant Breeding*. Blackwell Scientific Publications, Oxford.